Series/Number 07-126

LATENT CLASS
SCALING ANALYSIS

C. MITCHELL DAYTON
University of Maryland

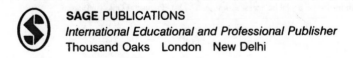

SAGE PUBLICATIONS
International Educational and Professional Publisher
Thousand Oaks   London   New Delhi

For information:

SAGE Publications, Inc.
2455 Teller Road
Thousand Oaks, California 91320
E-mail: order@sagepub.com

SAGE Publications Ltd.
6 Bonhill Street
London EC2A 4PU
United Kingdom

SAGE Publications India Pvt. Ltd.
M-32 Market
Greater Kailash I
New Delhi 110 048 India

Printed in the United States of America

Library of Congress Cataloging-in-Publication Data

Dayton, C. Mitchell (Chauncey Mitchell)
    Latent class scaling analysis / by C. Mitchell Dayton.
        p.   cm. — (A Sage university papers series. Quantitative
    applications in the social sciences : #07-126)
        Includes bibliographical references and index.
        ISBN 0-7619-1323-8 (pbk. : acid-free paper)
        1. Latent structure analysis.   I. Title.   II. Series: Sage
    university papers series. Quantitative applications in the social
    sciences : no. 07-126.
    QA278.6.D39   1999
    519.5'35—dc21   98-43854

13 14 15 16 17 7 6 5 4 3 2

Acquiring Editor:	C. Deborah Laughton
Editorial Assistant:	Eileen Carr
Production Editor:	Astrid Virding
Production Assistant:	Denise Santoyo
Typesetter:	Technical Typesetting Inc.

When citing a university paper, please use the proper form. Remember to cite the Sage University Paper series title and include paper number. One of the following formats can be adapted (depending on the style manual used):

(1) DAYTON, C. M. (1998) *Latent Class Scaling Analysis.* Sage University Papers Series on Quantitative Applications in the Social Sciences, 07-126. Thousand Oaks, CA: Sage.

(2) Dayton, C. M. (1998). *Latent class scaling analysis* (Sage University Papers Series on Quantitative Applications in the Social Sciences, series no. 07-126). Thousand Oaks, CA: Sage.

CONTENTS

Social science measures are observed, the concepts they measure are not. To measure "political ambition," say, a political scientist might administer a multiple item questionnaire to a group of party activitists and prospective candidates for office. The item scores somehow manifest the ambition variable, which is not seen directly, but is latent. How to model this latent structure? The choice depends on many things, among them the level of measurement for the manifest and latent variables. If both are continuous, then some factor analytic method might be appropriate. [In this series, see Kim and Mueller, *Introduction to Factor Analysis* (07-13) and *Factor Analysis* (07-14); Long, *Confirmatory Factor Analysis* (07-33); Dunteman, *Principal Components Analysis* (07-69).] However, suppose that both variables are categorical. Then latent class analysis, which is sometimes viewed as a sort of "factor analysis for nominal data," becomes preferred. [In this series, see McCutcheon, *Latent Class Analysis* (07-64).] What if the measurement level is ordinal? In our example, suppose "political ambition" is a scale ordered "low, medium, high." If the variables are so ordered, then latent class scaling models, carefully explicated here by Dr. Dayton, should be turned to.

The classic example of the kind of scale under study here is the Guttman scale, where the variable scores follow an order. For instance, in the political ambition survey, there might be three dichotomous variables, X, Y, and Z. With X, the respondent is asked, "Would you be willing to speak at local clubs"?; with Y, "Would you be willing to go on the road and make speeches?"; with Z, "Would you be willing to move to Washington?" The variables are ordered in terms of the degree if difficulty of the political commitment. Presumably, therefore, the respondent who said "yes" to Y would have said "yes" to X, and someone who said "yes" to Z would have had to say "yes" to X and Y. In summary, the theoretical response vectors are {000}, {100}, {110}, and {111}, where 1 = yes, 0 = no. Obviously, not all respondents will give a "theoretically correct" response.

For example, someone might say "yes" to Y and "no" to X and Z, i.e., {010}. Dr. Dayton works with the assumption that such an incorrect response is an error, which has the advantage of allowing a probabilistic treatment.

The latent class scaling models he focuses on apply to such binary response variables. The empirical examples are plentiful, and well chosen to teach the technique. Substantive illustrations include a survey on academic cheating, children's mastery of spatial tasks, medical diagnosis of lung disease, attitudes toward the Army, and behavior during role conflict. Furthermore, the computer programs for latent class analysis are carefully reviewed, and a Web site is offered for keeping abreast of the latest developments. This monography fully explicates the many modeling options available, which makes it of special value to researchers doing detailed scale analysis in psychology, sociology, and education.

—*Michael S. Lewis-Beck*
Series Editor

ACKNOWLEDGMENT

The authors wish to thank reviewers, Jacques Hagenaars, Tilburg University and Peter Marsden, Harvard University for the thoughtful critiques and valuable suggestions for improving this manuscript.

LATENT CLASS SCALING ANALYSIS

C. MITCHELL DAYTON
University of Maryland

1. INTRODUCTION AND OVERVIEW

In their classic work, *Latent Structure Analysis*, Paul Lazarsfeld and Neil Henry (1968) brought together in a single, unified treatise, the fundamental principles for specifying the "meaning" of social science concepts. Given the vagueness, or fuzziness, of concepts such as honesty, ability, anxiety, empathy, introversion, etc., they argue that manifest, behavioral indicators are linked to concepts by probability relations and not by rigid laws. Latent structure models have been developed for a wide range of applications. It is useful to conceptualize the nature of the manifest variables as either continuous or categorical and, similarly, to view the underlying concepts as representing continuous or categorical latent variables. At the risk of some oversimplification, factor analysis (FA), along with its generalized formulation, structural equation modeling (SEM), is concerned with manifest and latent variables that both represent continua, whereas the focus of latent class analysis (LCA) is on models where both the manifest and latent variables are categorical. Other types of latent structure models may represent combinations of continuous and categorical variables. For example, the psychological measurement models known as item response theory (IRT) often assume that the manifest variables (e.g., responses to achievement test items) are categorical, whereas the underlying latent variable (e.g., ability) is continuous. On the other hand, the statistical distributional models known as discrete mixture models may assume continuous manifest variables following, for example, normal distributions, whereas the underlying latent structure is categorical.

In recent decades, major advances in latent-class modeling have occurred, and many of these advances have resulted in practical applications as a result of the increasing availability of computational capability. Indeed, specialized applications have developed to the point

1

where it seems advisable to provide applied researchers with guides to specific types of applications. Thus, the current volume deals with latent class models for ordered scales, or hierarchies, that arise in many fields including sociology, psychology, medicine, business, and education. By "hierarchies" we mean situations in which the underlying latent class structure is assumed to obey some specified ordering properties. For example, the first of two latent classes might represent a consistently higher level of a latent variable conceptualized as "honesty." In particular, we consider research situations in which data are available for several dichotomous (or, more generally, polytomous) response variables. In addition, it is assumed that the latent structure consists of two or more discrete latent classes. In the simplest case of two classes, each manifest variable is related, in probability, to each of the latent classes. An example from the research literature (Dayton and Scheers, 1997) is based on a survey that requested yes or no responses to a set of questions related to academic dishonesty (e.g., "Have you ever cheated by copying answers from someone sitting near you?"). Latent class analysis conducted with these data suggests two types of respondents that may be characterized as persistent cheaters or as noncheaters. Persistent cheaters have higher probabilities of answering yes to each of the survey questions, although a persistent cheater may not, in fact, have engaged in all of the different cheating behaviors, at least not in the time frame specified by the survey. Similarly, a noncheater may have occasional lapses. However, the probability of a yes response for a noncheater is relatively low for each of the survey questions. Note that this conceptualization of the persistent cheater is probabilistic, or fuzzy, rather than deterministic.

In more complex cases, there may be a series of latent classes representing different levels of an ordered scale, or hierarchy, and, once again, it is assumed that there is a probability model connecting each manifest variable to these levels. For example, in the child development literature, there is evidence for a three-stage developmental sequence in children's mastery of right–left spatial tasks (Whitehouse, Dayton, and Eliot, 1980). Successful performance of a task such as "Show me your left hand" occurs at an earlier developmental stage than performance of a task such as "Put your left hand on your right knee," which involves cross-midline identification. This task, in turn, is performed at an earlier developmental stage than a task such as "Put your left hand on my right hand," which involves reverse opposite identification. However, manifest responses are probabilistically, rather than deterministically, related to these stages and, in particular, for this particular data set, several children performed the second task despite failure to perform the first task. In conceptualizing scales such as this, it is useful to define ideal types that can be represented by permissible response vectors. For the three right–left spatial tasks in the order presented, the ideal types would be children at developmental levels consistent with one of the response vectors {000}, {100}, {110}, or {111}, where 1 represents successful performance of the task and 0 represents failure to perform the task.

For the sake of consistency, some standard terminology will be used to describe special types of models throughout this monograph. If the latent structure is conceptualized as comprising two latent classes, as represented by the example of the cheating survey, the model is referred to as "extreme types." These models are labeled extreme types because the probability of a positive response is expected to be consistently higher for one of the two classes. In fact, the analysis may or may not reveal this consistent pattern, and the appropriateness of the extreme-types description must be judged empirically. In cases for which we assume simple prerequisite relationships among responses to the manifest variables, as was true for the right–left spatial example, the term "linear scale" will be used or, equivalently, the term "Guttman scale" may be applied in recognition of the pioneering work of Louis Guttman (1947). The distinctive characteristic of a linear, or Guttman, scale is that the variables are ordered in some specific fashion. The ordering often is based on a theoretical model or on previous empirical evidence. In some cases, an ordering may be proposed because of subsuming relationships among the variables. Consider, for example, addition/subtraction, multiplication, and division test items that are used to assess the acquisition of simple arithmetic skills by school children. An ordering can be derived from the fact that successful division involves both multiplication and addition/subtraction skills, whereas multiplication involves addition/subtraction skills. Thus, for a sample of students with varying backgrounds in these skills, one might expect successful performance of higher level skills to, necessarily, require success on lower level skills. Therefore, the ideal types for three items would be represented by the same permissible response vectors as for the right–left spatial example: {000}, {100}, {110}, and {111}. Note that the permissible response vector, {000}, is included because some of the children may not be able to perform any of the arithmetic tasks.

For some sets of variables, the prerequisite relationships may be more complex than can be represented by a single linear scale. Thus, we consider the possibility of biform and multiform scales for which two or more sets of dependency relationships exist. Airasian (1969), for example, presents achievement data for three test items on the topic of chemistry. Although, in the abstract, a linear scale seems reasonable for the data, in fact, there were a substantial number of students who responded correctly to the second item, but not to the first item. This suggests the possibility of a biform scale based on two different orderings of the items. Assuming that the items are ordered, say, ABC, then one set of ideal types is represented by the permissible response vectors {000}, {100}, {110}, and {111} as before, but the second set is {000}, {010}, {110}, and {111}. Note that item C is dependent on both items A and B for both ideal types, but that items A and B operate independently in the sense that neither is prerequisite to the other. The set of response vectors corresponding to ideal types for the biform scale is the union of those for the two separate ideal types: {000}, {100}, {010}, {110}, and {111}.

It can be argued that the modern treatment of linear scaling models began with Proctor's (1970) development of a probabilistic model based on the notion of ideal types. His formulation of the Guttman scaling model was, in fact, a restricted latent class model in which he assumed that non-scale-type responses arose due to response "errors." This notion was extended by Dayton and Macready (1976) to encompass more than one type of response error. In fact, both of these models, as well as a great variety of others, are subsumed by models presented by Goodman (1974). It should be noted that the latent class models presented here are limited to linear scales based on dichotomous-response variables. Similar models have been developed for rating scale items that use a Likert-type response format (e.g., Rost, 1985, 1988). In addition, alternative approaches to scaling can be based on latent trait theory (for example, see van der Linden and Hambleton, 1997), and some latent trait techniques are distribution-free (e.g., Mokken and Lewis, 1982; Sijtsma, 1988).

In the next chapter of this monograph, some background and theoretical development for latent class models is summarized. Although the estimation procedures are somewhat complex and require fairly sophisticated computing capability, the underlying concepts involve only elementary probability notions. For readers who wish to explore the mathematical subtleties of latent class modeling in more depth, two additional sources, in addition to Goodman (1974), that would prove useful are Haberman (1979) and Hagenaars (1990). For a broader treatment of latent class modeling that does not specifically focus on linear scaling, some useful references are McCutcheon (1987), Clogg (1995), and Dayton (1991).

2. LATENT CLASS MODELS

A. The General Model

We assume that the data available for analysis consist of observations from a sample of N individual cases for a set of V dichotomous variables. Although the models that we consider can be generalized to incorporate polytomous-response variables, the majority of analyses found in the research literature are based on dichotomies, and it greatly simplifies notation to maintain this restriction. In practice, it is not uncommon to transform categorical variables to dichotomies even though the original form of the data is polytomous. For example, if a survey allowed the responses "yes," "no," or "undecided," the latter two responses might be combined so that the new response categories represent agreement or lack of agreement with the survey item. Although there is certainly some loss of information when responses are combined, the practice may be justified on the basis of sharpening and simplifying the models as well as their interpretation (e.g., especially when a category such as "undecided" is infrequently selection by respondents).

Assume that the number of dichotomous variables is V and that the variables are labeled A, B, C, etc. Because the variables are dichotomous, responses can be encoded 1/0 and, depending upon the context, these response codes may stand for "agree/disagree," "correct/incorrect," "survived/died," "favorable/unfavorable," etc. Ordinarily, the response coded as 1, in some sense, represents the outcome of primary interest to the researcher.

For a set of V dichotomous variables, there are 2^V different patterns of responses that can be observed. For example, with $V = 3$ variables, the $2^3 = 8$ different response vectors are $\{000\}$, $\{100\}$, $\{010\}$, $\{110\}$, $\{001\}$, $\{101\}$, $\{011\}$, and $\{111\}$. Of course, with real data, some of these patterns may not actually occur. In general, the responses for a sample of N cases can be summarized in a frequency table that shows the 2^V response vectors along with the number of cases that display each pattern. For example, three chemistry test items, labeled A, B, and C and scored 1/0 for correct/incorrect responses, were presented by Airasian (1969) for $N = 70$ high school students (Table 2.1). Note that 14 students failed to answer any of the items correctly (i.e., response vector {000}), eight students answered all three items correctly (i.e., response vector {111}), and two response vectors, {001} and {011}, were not observed for any student.

In the remainder of this section, some fundamental theory of modern latent class analysis is summarized. This theory encompasses models for linear scales, but is more general in scope and is intended to introduce basic concepts that apply to many different applications of latent class models.

Mathematical Model

An unrestricted latent class model includes two or more latent classes, each of which has its own unique set of conditional probabilities for the V manifest variables. Each conditional probability represents the proportion of times that, within a given latent class, the response "1" (i.e., correct, agree, etc.) occurs for a variable. In presenting the formal model, the notation of Goodman (1974) is utilized and attention is restricted to the case of three manifest variables, A,

TABLE 2.1
Airasian Chemistry Test Items

Item {ABC}	Freq.
{000}	14
{100}	17
{010}	9
{110}	20
{001}	0
{101}	2
{011}	0
{111}	8
Total	70

B, and C, with levels $i = \{1, 0\}$, $j = \{1, 0\}$, and $k = \{1, 0\}$, respectively. Furthermore, let the latent variable be denoted X with levels $t = \{1, \ldots, T\}$, where T is the number of latent classes. Within each latent class, there are conditional probabilities associated with the responses to the manifest variables. For the tth latent class, these conditional probabilities are represented by $\pi_{it}^{\bar{A}X}$, $\pi_{jt}^{\bar{B}X}$, and $\pi_{kt}^{\bar{C}X}$, where the use of an overbar indicates conditionality. Thus, in the first latent class, the conditional probabilities for the variables A, B, and C are $\pi_{i1}^{\bar{A}X}$, $\pi_{j1}^{\bar{B}X}$, and $\pi_{k1}^{\bar{C}X}$; in the second latent class, they are $\pi_{i2}^{\bar{A}X}$, $\pi_{j2}^{\bar{B}X}$, and $\pi_{k2}^{\bar{C}X}$; and so forth. Explicitly, the conditional probability, $\pi_{i1}^{\bar{A}X}$, denotes the probability that manifest variable A takes on the value $i = \{1, 0\}$ **given that** the response arises from a member of latent class 1.

As is true for probabilities in general, the conditional probabilities are restricted to be between 0 and 1 (e.g., $0 \leq \pi_{it}^{\bar{A}X} \leq 1$) and the sum of the conditional probabilities across response levels is 1 (e.g., $\pi_{11}^{\bar{A}X} + \pi_{01}^{\bar{A}X} = 1$). In addition, it is assumed that the proportion of respondents in latent class t is equal to π_t^X. For example, for a model with two latent classes, the latent class proportions are denoted π_1^X and π_2^X with the conditions $0 \leq \pi_1^X \leq 1$, $0 \leq \pi_2^X \leq 1$, and $\pi_1^X + \pi_2^X = 1$.

Given the foregoing concepts and notation, and letting $\mathbf{y}_s = \{i, j, k\}$ be the vector of 1/0 responses to the variables for the sth case, a general, unrestricted latent class model can be formulated in two steps. First, for any response vector, $\mathbf{y}_s$, the probability **assuming membership in latent class t,** is:

$$P(\mathbf{y}_s \mid t) = \pi_{ijkt}^{\bar{A}\bar{B}\bar{C}X} = \pi_{it}^{\bar{A}X} \times \pi_{jt}^{\bar{B}X} \times \pi_{kt}^{\bar{C}X} \qquad (2.1)$$

Equation 2.1 is a product binomial based on conditional probabilities. Because $\pi_{1t}^{\bar{A}X}$ is the probability of a 1 response to variable A in latent class t, whereas $\pi_{0t}^{\bar{A}X} = 1 - \pi_{1t}^{\bar{A}X}$ is the probability of a 0 response to variable A in latent class t, Equation 2.1 includes $\pi_{1t}^{\bar{A}X}$ if $i = 1$, but includes $\pi_{0t}^{\bar{A}X}$ if $i = 0$. The model in Equation 2.1 embodies a fundamental theoretical concept, known as local independence, that underlies latent class analysis. That is, **the observed responses to the manifest variables are assumed to be independent given that latent class membership is taken into account.**

The second step in formulating the general model is to write the unconditional probability for a response vector as a weighted sum across latent classes:

$$P(\mathbf{y}_s) = \sum_{t=1}^{T} \pi_t^X \times \pi_{it}^{\bar{A}X} \times \pi_{jt}^{\bar{B}X} \times \pi_{kt}^{\bar{C}X} \qquad (2.2)$$

Note that the right-hand side of Equation 2.2 is of the form

$$\pi_1^X \times P(\mathbf{y}_s \mid t = 1) + \pi_2^X \times P(\mathbf{y}_s \mid t = 2) + \cdots + \pi_t^X \times P(\mathbf{y}_s \mid t = T)$$

That is, each conditional probability for the response vector is weighted by the corresponding latent class proportion, and the sum of these terms provides the overall, or unconditional, probability for the response vector. For any real data set, we do not know the values of either the latent class proportions, π_t^X, or the conditional probabilities, $\pi_{it}^{\bar{A}X}$, $\pi_{jt}^{\bar{B}X}$, or $\pi_{kt}^{\bar{C}X}$. Typically, even the number of latent classes, T, is not known with any certainty prior to carrying out an analysis. In fact, the problem is even more complex than this because we may think of the latent variable, X, as representing combinations of more fundamental latent variables, say X_1 and X_2, that are subject to certain restrictions. For example, if X is at $T = 4$ levels, these levels may represent the combinations of two fundamental, uncorrelated dichotomous latent variables (see Hagenaars, 1990; such models are beyond the scope of the current volume). Thus, the problem in latent class analysis is similar to that in factor analysis. That is, we must decide among models with varying numbers of latent variables, each with varying numbers of classes, and must estimate the conditional probabilities associated with each model. Also, as will be developed in later sections of this monograph, many types of scaling models are based on restricted versions of the general model summarized here.

Predicting Latent Class Membership

An additional important component of latent class analysis is predicting latent class membership for cases showing various observed response vectors. Using Bayes' theorem, the posterior probability of membership in latent class t, given the response vector, $\mathbf{y}_s$, is

$$P(t \mid \mathbf{y}_s) = \frac{P(\mathbf{y}_s \mid t) \times \pi_t^X}{\sum_{t=1}^{T} P(\mathbf{y}_s \mid t) \times \pi_t^X} \qquad (2.3)$$

The first term in the numerator, $P(\mathbf{y}_s \mid t)$, is given by Equation 2.1, whereas the second term, π_t^X, in the numerator is the latent class proportion. Note that the latent class proportion plays the role of a prior probability in this formula and, also, note that the denominator is simply the unconditional probability for the response vector as given in Equation 2.2. The actual classification procedure for a given response vector is to compute Bayes' theorem for each latent class and then classify all cases with the given response vector into the latent class for which the posterior probability is largest. When there are only two latent classes, it is convenient to compute the posterior odds for, say, the first latent class, odds $= P(t = 1 \mid \mathbf{y}_s)/P(t = 2 \mid \mathbf{y}_s)$, and then classify the cases in the first latent class if the odds are greater than 1.0 or in the second latent class otherwise.

The success of the classification procedure can be assessed by computing an estimated proportion of correct classifications based on the modal (i.e., largest) latent class for each response vector and the frequencies, n_s, for the response vectors. Then the proportion correctly classified, P_c, is defined as

$$P_c = \frac{\sum_{s=1}^{2^I} n_s \times \max[P(t \mid \mathbf{y}_s)]}{N} \qquad (2.4)$$

where $\max[P(t \mid \mathbf{y}_s)]$ is the posterior probability for the modal latent class for the response vector $\mathbf{y}_s$. When interpreting this proportion, it must be remembered that there is a chance level of correct classification that would be maximized by simply classifying all cases into whichever class had the largest latent class proportion. To correct for this, a statistic, λ (i.e., lambda; Goodman and Kruskall, 1954), can be defined. Letting π_M^X be the largest of π_1^X, π_2^X, etc., the λ statistic is defined as

$$\lambda = \frac{P_c - \pi_M^X}{1 - \pi_M^X} \qquad (2.5)$$

In practice, the true values of π_1^X, π_2^X, etc. are unknown, but are replaced by data-based estimates as described in Section 2.B.

It should be noted that the success of classification as indexed by P_c or λ for an actual data set tends to be optimistic, because both parameter estimation and classification are based on the same data. The magnitude of the upward bias and methods to correct for it have not been studied in any detail in the context of latent class analysis, but the bias is not likely to be substantial for large data sets.

Hypothetical Example

An example is useful to clarify the concepts presented so far. The hypothetical data presented in Table 2.2 are for $N = 1,000$ respondents who answered two agree/disagree survey items, labeled A and B, dealing with their interest in artistic topics. Note that $144 + 176 = 320$, or 32%, of the respondents answered "agree" (i.e., "yes") to item A, whereas $164 + 176 = 340$, or 34%, of the respondents answered "agree" to item B. If, in the sample of 1,000 respondents, the responses to the two items were independent, about 109 individuals would have agreed with both items (i.e., $1,000 \times .32 \times .34 = 108.8 \simeq 109$). In fact, 176 individuals are shown as agreeing with both items, and this suggests lack of independence for responses to the two items (more formally, the Pearson chi-square statistic for the test of independence is 91.1 with 1 degree of freedom, and this value is significant at any conventional level).

Despite this apparent lack of independence, the data in Table 2.2 were generated by positing two latent classes of respondents, and responses to the two items **within** each class were independent. The first type, or class, of respondents had relatively high levels of artistic interest and the probabilities of agreeing with items A and B were .8 and .9, respectively (i.e., $\pi_{11}^{\bar{A}X} = .80$, $\pi_{11}^{BX} = .90$). On the other hand, the second type of respondents, who had relatively lower levels of artistic interest, had probability of agreeing of .2 for both items A and B (i.e., $\pi_{12}^{\bar{A}X} = \pi_{12}^{BX} = .20$). Furthermore, the sample contained 20% of the more artistic types (i.e., $\pi_1^X = .20$) and 80%

TABLE 2.2
Hypothetical Data for Two Items

Item {AB}	Freq.
{00}	516
{10}	144
{01}	164
{11}	176
Total	1,000

of the less artistic types (i.e., $\pi_2^X = .80$). Table 2.3 displays proportions and frequencies separately for the two latent classes. For example, the proportion in the first latent class that shows the response $\{10\}$ is $.2 \times .8 \times (1 - .9) = .016$, whereas the proportion in the second latent class that shows this response is $.8 \times .2 \times (1 - .2) = .128$. The frequencies are simply the proportions multiplied by the sample size, 1,000. Given the frequencies in each latent class, it is easy to verify that the responses are independent **within latent classes.** For example, note that for the artistic latent class, 90% of those responding "1" to item A and 90% of those responding "0" to item A, respond "1" to item B (i.e., the proportions are $144/160 = .9$ and $36/40 = .9$, respectively). Thus, the apparent lack of independence of the responses in Table 2.2 was the result of mixing two different types of respondents—those with and those without high levels of artistic interest—but within each of these classes, the condition of local independence is satisfied.[1] With real data, of course, a latent class model cannot be expected to fit the data perfectly as was the case for these hypothetical data. However, as subsequently discussed, there are some useful approaches to deciding whether or not a given model represents a reasonable fit to the data.

The results of applying Bayes' theorem, Equation 2.3, to these hypothetical data are displayed in Table 2.4. The Bayes probabilities can be computed directly from the entries in Table 2.3. For example, $P(t = 1 \mid \mathbf{y}_s = \{10\}) = .016/(.016 + .128) = .111$. Note that only for the response vector $\{11\}$ would an individual be classified as belonging to the higher artistic latent class, because it is only for this response vector that the posterior probability is larger than for the nonartistic

TABLE 2.3
Latent Class Structure for Two Hypothetical Items

Item {A, B}	Freq.	Proportions		Frequencies	
		Latent Class 1	Latent Class 2	Latent Class 1	Latent Class 2
{00}	516	0.004	0.512	4	512
{10}	144	0.016	0.128	16	128
{01}	164	0.036	0.128	36	128
{11}	176	0.144	0.032	144	32
Total:	1,000	0.200	0.800	200	800

TABLE 2.4

Bayes Classifications for Hypothetical Data

Item {AB}	Freq.	$P(c = 1 \mid Y)$	$P(c = 2 \mid Y)$	Class
{00}	516	0.008	0.992	2
{10}	144	0.111	0.889	2
{01}	164	0.220	0.780	2
{11}	176	0.818	0.182	1
Total	1,000			

class. The proportion correctly classified, P_c, is equal to $[(516 \times .992) + (144 \times .889) + (164 \times .780) + (176 \times .818)]/1,000 = .912$ and, correcting for chance, the lambda statistic is $\lambda = (.912 - .800)/(1 - .800) = .56$. That is, applying Bayes' theorem for purposes of classification led to a 56% improvement over the chance proportion of correct classifications that is obtained by placing everyone in the nonartistic class.

B. Estimating Parameters

Maximum-Likelihood Estimates

Whereas latent class models fall into the category of log-linear models with latent variables, maximum likelihood is the conventional approach to parameter estimation. Using the definition in Equation 2.2, the likelihood for a sample of N cases is the product

$$L = \prod_{s=1}^{2^I} P(\mathbf{y}_s)^{n_s}$$

where n_s is the observed frequency of cases for the response vector $\mathbf{y}_s$. In effect, the likelihood is the probability, given the model, for the ordered sample of cases. The maximum-likelihood estimates (MLEs) for the latent class proportions, $\hat{\pi}_t^X$, and conditional probabilities, $\hat{\pi}_{it}^{\bar{A}X}$, $\hat{\pi}_{jt}^{\bar{B}X}$, and $\hat{\pi}_{kt}^{\bar{C}X}$, for some specified latent class model are found by maximizing L with respect to the parameters [in general, the maximum-likelihood estimate for a parameter in a model is indicated by a caret (ˆ)]. Unfortunately, unlike widely known statistical procedures such as regression analysis and analysis of variance, there are no general-purpose closed-form formulae for computing these MLEs. However, iterative procedures, such as the estimation-maximization (EM) algorithm or Fisher's method of scoring, among others, can be used to find the MLEs as long as certain identifiability conditions are met by the latent class model under consideration.

Model Identification

From a mathematical perspective, the sufficient condition for identifiability of a latent class model is that the theoretical covariance matrix for the maximum-likelihood estimates be positive-definite. This requirement is somewhat technical, but means that there is no collinearity among the parameters. A second condition that is necessary, but not sufficient, is simply that there be zero or positive degrees of freedom for the model. For V dichotomous variables, there are 2^V unique response vectors with associated frequencies of occurrence. The degrees of freedom for the model are $2^V - m - 1$, where m is the total number of independent parameters in the latent-class model. For example, for an unrestricted two-class model, there are, in general, $2V + 1$ independent parameters. Thus, for $V = 3$, there are $8 - 7 - 1 = 0$ degrees of freedom, and this is the minimum number of variables for which an unrestricted two-class model can be estimated (in particular, an unrestricted two-class model cannot be estimated for only two variables). It should be noted, however, that nonnegative degrees of freedom do not necessarily imply that a model is identified. For example, the unrestricted three-class model with $V = 4$ variables is not identified (i.e., the covariance matrix for the maximum-likelihood estimates is not of full rank) even though the degrees of freedom are $16 - 12 - 2 - 1 = 1$. Additional details concerning estimation with the EM algorithm and identifiability are summarized in McCutcheon (1987, p. 21–27) and in Bartholomew (1987, Chap. 2).

Computer Programs for Latent Class Analysis

One of the oldest and most widely available programs for conducting latent class analysis is MLLSA by Clogg (1977). This program uses the EM algorithm for computing purposes and has the important advantage that it provides for the assessment of the rank of the estimated covariance matrix and, thus, aids in determining the identifiability of models. The original version of MLLSA was written for mainframe computers, but several microcomputer adaptations have been produced.[2] A brief description and some sample runs are summarized on the Web site, //www.inform.umd.edu/EDUC/Depts/EDMS. Another popular latent class program, LCAG (Hagenaars and Luijkx, 1987), is also based on the EM algorithm and has some features that simplify fitting certain advanced models. Also, LCAG was designed to allow analyses that include cases with partial missing data by permitting the input of subtables (e.g., for cases with missing observations on variable D, data can be inputted for the $A \times B \times C$ subtable). Finally, the program LEM by Vermunt (1993) handles a wide variety of models including latent class models and has the advantage that it provides estimated standard errors for the parameters estimates. Unlike MLLSA or LCAG, latent class models in LEM also can be conceptualized as log-linear models with latent variables (see Section 2.D.). LEM incorporates most, if not all, of the features of MLLSA and LCAG, and has the option of inputting raw data rather than a table of frequencies for the observed response vectors (this is useful when the number of manifest variables is large). Although many of the analyses reported in this book were conducted using MLLSA, LEM has a modern computer interface and is available on the Web (see the Web site //cwis.kub.nl/~fsw_1/mto/).

An interesting characteristic of programs based on the EM algorithm is that parameter estimates are computed even when the model being fitted to the data is not identified. As noted, MLLSA, as an option, does provide an estimate of the rank of the covariance matrix for the maximum-likelihood estimates and provides the appropriate degrees of freedom for assessing fit of a suitably restricted, identified model. However, when a model is not identified, there are, in general, many different sets of parameter estimates that fit the data equally well and, in arriving at a solution, programs like MLLSA impose some implicit restrictions on the solution for the parameter estimates. Additional discussion of the identification issue and its implications for computation can be found in Clogg (1995).

C. Assessing the Fit of Models to Data

The data for the hypothetical example in Section A were generated from a specific, known latent class structure. With real-world data sets, the latent structure is unknown and it is the analyst's task to estimate the parameters of the latent class model of interest and to decide whether or not the given model fits the observed data within acceptable limits. In practice, there are often several competing models of interest that must be compared. There are several techniques that can be of aid in this endeavor, and three of these are developed in this section: chi-square significance tests, measures of relative fit based on information criteria, and indices of model fit, including an index, π^*, developed by Rudas, Clogg, and Lindsay (1994).

Chi-Square Goodness-of-Fit Tests

The goodness of fit of a given latent class model to the observed data can be assessed using a chi-square test based on observed and expected frequencies. There are two versions of the chi-square statistic in common use: the familiar Pearson statistic, X^2, which is based on the differences between observed and expected frequencies, and the less familiar likelihood-ratio, G^2, statistic, which is based on the logarithm of the ratio of observed and expected frequencies. Assuming that maximum-likelihood parameter estimates have been computed, expected probabilities, $\hat{P}(y_s)$, are found by substituting into Equation 2.2 for each of the 2^V response vectors. Then, based on a total of N observations, the expected frequency for the sth response vector is

$$\hat{n}_s = N \times \hat{P}(y_s) \qquad (2.6)$$

Letting n_s be the corresponding observed frequency for the response vector, the two chi-square statistics are

$$X^2 = \sum_{s=1}^{2^V} \frac{(n_s - \hat{n}_s)^2}{\hat{n}_s} \qquad (2.7)$$

$$G^2 = 2 \sum_{s=1}^{2^V} n_s \times \log_e\left(\frac{n_s}{\hat{n}_s}\right) \qquad (2.8)$$

When $n_s = 0$, the corresponding term in G^2 is set equal to 0 [i.e., by convention, $0 \times \log_e(0) = 0$]. In theory, the degrees of freedom for both of these chi-square statistics are equal to the number of unique response vectors (e.g., for dichotomous data, usually 2^V) minus one more than the number of independent parameters estimated; that is, $2^V - m - 1$, where m is the number of independent parameters. However, if the rank of the asymptotic covariance matrix is less than this amount, then the degrees of freedom for a suitably restricted, identified model must be adjusted accordingly. For example, as noted earlier, with four variables and three latent classes, the apparent number of independent parameters, m, is equal to 14, but the rank of the asymptotic covariance matrix is only 13. Hence, the degrees of freedom for fitting an identified model are actually $16 - 13 - 1 = 2$.

As previously noted when discussing computer programs for latent class analysis, when a latent class model is not identified, there are, in general, many different restricted models that will result in exactly the same set of expected frequencies, $\hat{n}_s$, and, hence, in exactly the same value for a chi-square fit statistic. Thus, the interpretation of the parameter estimates from such models is not meaningful. The arbitrariness of parameter estimates for models that are not identified can be demonstrated easily by changing the start values for the solution in a program such as MLLSA. Changing start values for either or both the latent class proportions and the conditional probabilities will, in general, alter the estimates (sometimes dramatically), but will not change the chi-square fit statistics.

Expert opinion is divided concerning which chi-square statistic, X^2 or G^2, is better to use in practice to assess the fit of models. If there are small expected frequencies (e.g., expected frequencies less than 1.0 are always considered small; some researchers consider 5.0 small, however), both statistics tend to be distorted and may fail to follow the appropriate theoretical chi-square distribution. If there is a large difference between the computed values of X^2 and G^2, there are often one or two response vectors where small expected frequencies are causing the discrepancy. Combining response vectors with small expected frequencies and adjusting degrees of freedom accordingly (i.e., use the total number of response vectors after combining rather than 2^V in the formula for degrees of freedom) will alleviate this problem in some applications.

The problem of small expected frequencies is discussed by Read and Cressie (1988), who present a generalized version of the chi-square statistic in the so-called power-divergence family of goodness-of-fit statistics. The general form of the Read–Cressie statistic for V dichotomous variables is

$$I^2 = \frac{2}{\lambda(\lambda+1)} \sum_{s=1}^{2^V} n_s \times \left[\left(\frac{n_s}{\hat{n}_s}\right)^{\lambda} - 1 \right] \qquad (2.9)$$

where the choice of λ determines the specific form of the statistic. In particular, the choice $\lambda = 0$ yields (by taking limits) the G^2 statistic of Equation 2.8 and the choice $\lambda = 1$ yields (by doing algebra) the X^2 statistic of Equation 2.7. For a variety of reasons, Read and Cressie (1988, Chap. 6) recommend the choice of $\lambda = 2/3$. There is little known about the performance of this recommendation in the context of latent class modeling, although the program LEM does compute the power-divergence statistic with $\lambda = 2/3$ (reported as Cressie–Read on the output). In Chapters 6 and 7, we report the Read–Cressie statistic for cases in which the frequency table is sparse (i.e., there are many 0 or small observed frequencies).

Chi-Square Difference Tests

In certain circumstances for nested models, it is appropriate to use chi-square tests for the purpose of comparing alternate models. Nested models exist when the model with fewer parameters can be derived by placing one or more restrictions on the model with the larger number of parameters. For example, a two-class model for four variables, A, B, C, and D, may be restricted by assuming that, within each latent class, all of the conditional probabilities are the same. Technically, these restrictions are $\pi_{11}^{\bar{A}X} = \pi_{11}^{\bar{B}X} = \pi_{11}^{\bar{C}X} = \pi_{11}^{\bar{D}X}$ and $\pi_{12}^{\bar{A}X} = \pi_{12}^{\bar{B}X} = \pi_{12}^{\bar{C}X} = \pi_{12}^{\bar{D}X}$. Then, the respective degrees of freedom for the unrestricted and restricted models are $16 - 9 - 1 = 6$ and $16 - 3 - 1 = 12$ (note that six restrictions have been imposed on the conditional probabilities and that the difference in degrees of freedom is equal to 6). For nested models, it **may** be appropriate to compare the two models as follows: Compute the difference between the goodness-of-fit chi-square statistics (in theory, the likelihood ratio chi-square statistic, G^2, should be used) for the two models and evaluate this difference with degrees of freedom equal to the difference in degrees of freedom between the two models (or equivalently, equal to the number of independent restrictions imposed).

Note that there are deliberate caveats in the description of chi-square difference tests. First, difference chi-square tests are not appropriate for comparing models based on different numbers of latent classes (e.g., for comparing a three-class to a four-class model) even though these models are nested. This limitation arises for technical reasons related to the fact that the nesting relationship between the models involves restricting parameters to their upper/lower limits of

0/1 that represent boundary values (see Everitt and Hand, 1981, for more technical details). An appreciation for this issue can be gained by considering the following case. An unrestricted three-class model can be derived from an unrestricted four-class model in two equivalent ways: (1) set $\pi_4^X = 0$ (note, then, that the values of $\pi_{i4}^{\bar{A}X}$, $\pi_{j4}^{\bar{B}X}$, etc., are irrelevant), or (2) set $\pi_{13}^{\bar{A}X} = \pi_{14}^{\bar{A}X}$, $\pi_{13}^{\bar{B}X} = \pi_{14}^{\bar{B}X}$, $\pi_{13}^{\bar{A}X} = \pi_{14}^{\bar{A}X}$, and $\pi_{13}^{\bar{A}X} = \pi_{14}^{\bar{A}X}$. The first approach involves one restriction, whereas the second approach involves four restrictions and, thus, the degrees of freedom for the difference chi-square statistic are ambiguous. A second caveat is that, in theory, the difference chi-square statistic for two nested models follows the appropriate, theoretical chi-square distribution only if the unrestricted model is a true model. The impact of violating this assumption has not been investigated to any great extent and, therefore, its importance cannot be evaluated. Because of these technical issues surrounding the use of chi-square difference tests, we recommend, in general, using the next two categories of procedures for making comparisons among models.

Information Criteria

The Akaike (1973, 1987) information criterion, AIC, is a measure of model fit based on concepts derived from information theory. For comparison of two or more different models that have been fitted to the same data, the AIC favors the model that is expected to show the smallest decrease in likelihood if the model were cross-validated on a new sample of cases. An important characteristic of the AIC is that the models under consideration need not be nested and, hence, the AIC provides a decision procedure that is more general than the use of chi-square difference statistics.

There are two equivalent ways to implement the AIC for a set of competing latent class models. As originally proposed by Akaike, the AIC is based on the logarithms of the likelihoods for the data based on, say, H different models being compared. If $P_h(\mathbf{Y}_s)$ is the probability for an observed response vector based on maximum-likelihood estimates for the hth model, as defined by Equation 2.2, then

$$\log_e(L_h) = \sum_{s=1}^{2^{I'}} n_s \times \log_e[P_h(\mathbf{Y}_s)],$$

$$\text{AIC}_h = -2\log_e(L_h) + 2m_h$$

(2.10a)

where m_h is the number of independent parameters that are estimated when fitting the hth model to the data. The Akaike decision procedure entails selecting the model with the minimum value of AIC as the preferred model. An alternate approach that is often simpler to use is based on likelihood-ratio chi-square statistics. For H different latent class models fitted to the same data, let G_h^2 be the chi-square value, as defined by Equation 2.8, and let ν_h be the corresponding degrees of freedom. Then, an alternative definition of the Akaike information criterion for the hth model is

$$\text{AIC}_h^* = G_h^2 - 2\nu_h \qquad (2.10\text{b})$$

It can be shown that AIC and AIC* differ only by a constant amount that involves the sample size, N, and the number of observed response vectors, 2^V, and, thus, a decision based on selecting the minimum of AIC_h or of AIC_h^* always results in the choice of the same preferred model. Note that for two models with similar values of $\log_e(L_h)$, the smaller and, therefore, favored value of AIC is associated with the model based on fewer parameters. Thus, the AIC is sometimes interpreted as a penalized version of the log likelihood, where the penalty term, $2\nu_h$, is intended to avoid overparameterization.

A criticism of the AIC is that it lacks properties of asymptotic (i.e., large sample) consistency because the definition of the AIC does not directly involve the sample size, N. Schwarz (1978) utilized a Bayesian argument to develop an asymptotically consistent measure [Bayesian information criterion (BIC)]:

$$\text{BIC}_h = -2\log_e(L_h) + \log_e(N) \times m_h \qquad (2.11)$$

An alternative version of the BIC can be computed using G^2 as in the case for AIC [i.e., $\text{BIC}_h^* = G_h^2 - \log_e(N) \times \nu_h$]. The model selection procedures for the BIC are the same as for the AIC. Whereas $\log_e(8) = 2.08$, the penalty term for BIC is larger than for AIC for sample sizes of 8 or more. Thus, as a rule, the BIC tends to select less complex models (i.e., models based on fewer parameters) than does the AIC. Results of empirical investigations of the AIC and BIC (Lin and Dayton, 1997) suggest that, in the context of latent class models, researchers might prefer the AIC unless the sample contains several thousand cases or the models being estimated are based on relatively few parameters, in which case the BIC is preferable. However, in other contexts such as time-series analysis, simulation results tend to favor the BIC (for a technical discussion of the AIC versus BIC, see Kass and Raftery, 1995). Often the two measures select the same or very similar models, and our strategy is to compute both the AIC and BIC for applications.

Indices of Model Fit

When fitting a model based on a large number of cases, hypothesis tests tend to result in rejection of the model even though, from a substantive point of view (e.g., in terms of residuals), the lack of fit of the model does not seem unsatisfactory to the researcher. This state of affairs resulted in the development of various fit indices, two of which are described in this section.[3]

For a latent class model fitted to V dichotomous variables, the index of dissimilarity, I_D, is defined in terms of observed and expected frequencies (i.e., Equation 2.4) as

$$I_D = \frac{\sum_{s=1}^{2^V} |n_s - \hat{n}_s|}{2N} \tag{2.12}$$

Although it is apparent that the minimum possible value of I_D is 0, the upper limit is less than 1 and varies for different types of models. As a rule, values of I_D less than .05 or so are considered small, although the actual interpretation of I_D should be with reference to some specific area of application.

Rudas, Clogg, and Lindsay (1994) proposed an index of lack of fit based on the notion of deleting cases from the observed frequency table such that the remaining cases fit the model of interest perfectly. Their index, π^*, is defined as the minimum proportion of cases that must be deleted to achieve perfect model fit. As an example, consider the data in Table 2.5 where we assume that the purpose is to fit an independence model. Note that if one case were removed from cell {00} and another case were removed from cell {11}, the remaining frequencies (labeled "Freq.*") would be exactly consistent with the hypothesis of independence (e.g., the odds in favor of $B = 1$ are 2:1 for both $A = 1$ and $A = 2$). Thus, the value of π^* for this data set is $2/47 = .043$. As with the index of dissimilarity, π^* must be interpreted within some practical context. For realistic data sets, the

TABLE 2.5
Hypothetical Data to Illustrate π^*

Item {AB}	Freq.	Freq.*
{00}	11	10
{10}	5	5
{01}	20	20
{11}	11	10
Total	47	45

computation of π^* is somewhat involved because the parameters in the model of interest (e.g., a two-class latent class model) must be estimated simultaneously with the estimation of π^*. In the next section of this monograph, we present a practical approach for computing π^* for latent class models that can often be implemented using microcomputer spreadsheet programs (e.g., Microsoft Excel).

Model of Complete Independence

If there is, in fact, no evidence for any association among the observed variables being analyzed, it is, in general, of no interest to undertake additional latent class analysis. Thus, as a preliminary step in all analyses, it is recommended that a model of complete independence be assessed for the frequency data. This model is equivalent to a one-class "latent class" model based on fitting expected frequencies using the marginal proportions for the manifest variables. The expected frequency for the response vector $\{i, j, k\}$ is of the form $\hat{n}_s = N \times \hat{\pi}_i^A \times \hat{\pi}_j^B \times \hat{\pi}_k^C$, where $\hat{\pi}_i^A$ is the proportion of cases that show the response i for manifest variable A, etc. The degrees of freedom for the chi-square test of complete independence for dichotomous manifest variables are always $2^V - V - 1$ (this test is available as output from the MLLSA program). More generally, for V manifest variables that may be polytomous, the degrees of freedom for the test of complete independence become $\prod_{v=1}^{V} K_v - \sum_{v=1}^{V} K_V + V - 1$, where K_v is the number of levels for the vth manifest variable.

Standard Errors

As previously discussed, two microcomputer programs, MLLSA and LCAG, are widely available for unrestricted and restricted latent class analysis. These programs are easy to use, but both have the disadvantage that they do not provide estimates for the standard errors of the latent class proportions or conditional probabilities. In this section, we describe two methods for empirically estimating standard errors that can be used in connection with these programs. For users of LEM, asymptotic estimates of standard errors are available, but their accuracy is unknown, especially for small sample sizes, and the procedures described in this section can be used as a check.

There are two major uses for estimated standard errors within latent class modeling. First, standard errors can be used to construct confidence intervals that give some sense of the stability of parameter estimates. For example, with large samples, a 90% confidence interval for the estimated value of the proportion in the first latent class would be of the form $\hat{\pi}_1^X \pm 1.645\hat{\sigma}_e$, where $\hat{\sigma}_e$ is some estimate for the standard error of $\hat{\pi}_1^X$. Note that for confidence coefficients other than .90, the multiplier 1.645 would be replaced by the appropriate value of the standard normal distribution (e.g., 1.96 for a 95% confidence interval). Second, various tests of significance can be constructed by using estimated standard errors. In particular, any given parametric value can be hypothesized to equal 0 and a large-sample z test set up. For example, to test the hypothesis, $H_0: \pi_{11}^{\bar{A}X} = 0$, the z test would be of the form $z = \hat{\pi}_{11}^{\bar{A}X}/\hat{\sigma}_b$, where $\hat{\sigma}_b$ is some estimate for the standard error of $\hat{\pi}_{11}^{\bar{A}X}$.

(a) Jackknife. The jackknife is a resampling method that is based on the notion of omitting one observation at a time and then recomputing the statistic of interest. Assume that, for a sample of N observations, the computed value of the statistic of interest based on the total sample is W and that, if the ith case is omitted, the computed value of the statistic based on the remaining $N-1$ cases is W_i, $i = 1, \ldots, N$ (note that temporarily we are using the subscript i to refer to a case rather than to a level of manifest variable A). Then the jackknife estimate for the sampling variance of W is $N \times \sum_{i=1}^{N}(W_i - W)^2/(N-1)$ and the square root of this quantity is the estimated standard error of W. Note that, for large N, the sampling variance of W is essentially the sum of squares of the jackknife estimates, W_i. With frequency data, omitting any particular case from a given cell (i.e., response vector) is the same as omitting any other. Thus, the jackknife computation for the statistic can be done once per cell after reducing the cell fre-

quency, n_s, by 1. Then the statistic computed from the reduced sample, W_s, is weighted by the frequency, n_s, in the computation of the jackknife standard error. Assuming a total of M cells, the jackknife estimate for the standard error using frequency data is

$$SE_J = \sqrt{\frac{N \times \sum_{s=1}^{M} n_s (W_s - W)^2}{N - 1}} \qquad (2.13)$$

To implement the jackknife using a program such as MLLSA, it is necessary to set up a series of analyses each based on reducing one of the cell frequencies by 1. For example, with four dichotomous variables and assuming that all 16 of the cell frequencies, n_s, are nonzero, it would be necessary to set up a series of $M = 2^4 = 16$ analyses. The remaining computations can be done in a spreadsheet. This approach is illustrated in the next chapter.

(b) Parametric Bootstrap. Another resampling technique, the parametric bootstrap, involves, in essence, conducting a simulation study in which the sample values of the latent class proportions and conditional probabilities are treated as if they were population values. Random samples of size N are generated using these population values, the statistic of interest is computed from the samples, and the standard deviation of the empirical distribution is used to estimate the standard error of the statistic. In general, the bootstrap requires some specialized programming, but for readers who have access to the matrix language Gauss, we have included programming code in the Web site, //www.inform.umd.edu/EDUC/Depts/EDMS, that will conduct the simulation for the case of an unrestricted two-class model with dichotomous response variables. As for the jackknife, the parametric bootstrap technique is illustrated in the next chapter.

For additional information concerning resampling techniques such as the bootstrap and jackknife, the article by Efron and Gong (1983), and the monograph by Mooney and Duval (1993) are recommended. Note that in the Efron and Gong reference, the description of sampling from the "parametric maximum likelihood distribution" is equivalent to our parametric bootstrap.

D. A Note on Notation

In this book, we utilize a probabilistic notation for latent class analysis (e.g., Equations 2.1 and 2.2) that was popularized by Goodman (1974) and that is used in documentation for the computer programs MLLSA (Clogg, 1977) and LCAG (Hagenaars and Luijkx, 1987). An alternative notation, based on the concepts of log-linear models, was introduced by Haberman (1979) and is used in the documentation for some other computer programs such as NEWTON (Haberman, 1988) and LEM (Vermunt, 1993). In this section, we summarize this notation for those readers who already have some familiarity with log-linear modeling.

Consider the cross-tabulation of three manifest variables, A, B, and C. A saturated log-linear model (**without** latent variables) for the logarithm of the expected frequency, $\hat{n}_{ijk}$, in cell $\{ijk\}$ is

$$\log_e(\hat{n}_{ijk}) = \lambda_0 + \lambda_i^A + \lambda_j^B + \lambda_k^C + \lambda_{ij}^{AB} + \lambda_{ik}^{AC} + \lambda_{jk}^{BC} + \lambda_{ijk}^{ABC} \quad (2.14)$$

where λ_0 is a constant, the terms λ_i^A, λ_j^B, and λ_k^C are effects for the manifest variables, the terms λ_{ij}^{AB}, λ_{ik}^{AC}, and λ_{jk}^{BC} are first-order interactions among the variables, and the term λ_{ijk}^{ABC} is the second-order interaction among all three variables. As in analysis of variance, these terms are constrained to sum to zero by rows, columns, etc. Thus, for dichotomous variables, $\sum_{i=1}^{2} \lambda_i^A = \sum_{j=1}^{2} \lambda_j^B = \sum_{k=1}^{2} \lambda_k^C = 0$; $\sum_{i=1}^{2} \lambda_{ij}^{AB} = 0$ for all j; $\sum_{j=1}^{2} \lambda_{ij}^{AB} = 0$ for all i; etc. If it is assumed that the variables are independent of one another, then all interaction terms are zero and the log-linear model reduces to

$$\log_e(\hat{n}_{ijk}) = \lambda_0 + \lambda_i^A + \lambda_j^B + \lambda_k^C \quad (2.15)$$

In the case of a latent class model, the manifest variables are assumed to be (conditionally) independent when the latent variable, X, is introduced. Thus, the log-linear model for the (unobserved) expected value for the cell $\{ijkt\}$ becomes

$$\log_e(\hat{n}_{ijkt}) = \lambda_0 + \lambda_i^A + \lambda_j^B + \lambda_k^C + \lambda_t^X + \lambda_{it}^{AX} + \lambda_{jt}^{BX} + \lambda_{kt}^{CX} \quad (2.16)$$

Note that this model contains an effect for the latent variable, X, and first-order interactions between X and the manifest variables,

A, *B*, and *C*, but that all other interaction terms are assumed to be zero. Setting these interaction terms to zero is equivalent to assuming local independence. The latent variable log-linear model in Equation 2.16 is for an expected frequency in the unobserved, "complete" data table that includes the observed variables, *A*, *B*, and *C* as well as the latent variable, *X*. Parameter estimates for this latent variable log-linear model can be obtained, for example, from the program LEM (although the constant term λ_0 is not included in the output and the estimates are labeled "beta"). Given estimates for the λ terms in Equation 2.16, it is possible to transform these estimates into latent class proportions and conditional probabilities corresponding to those in Equation 2.2, and the program LEM yields both sets of values. The transformation from logarithms of expected frequencies to probabilities is logistic in form and some examples are

$$\pi_1^X = \frac{\exp(2\lambda_1^X)}{1 + \exp(2\lambda_1^X)} \qquad \pi_{11}^{\bar{A}X} = \frac{\exp(2\lambda_1^A + 2\lambda_{11}^{AX})}{1 + \exp(2\lambda_1^A + 2\lambda_{11}^{AX})}$$

$$\pi_{12}^{\bar{A}X} = \frac{\exp(2\lambda_1^A - 2\lambda_{11}^{AX})}{1 + \exp(2\lambda_1^A - 2\lambda_{11}^{AX})}$$

3. EXTREME-TYPES MODELS

A. Saturated Models

A relatively simple, but very useful, scaling model posits just two latent groups. To qualify as a scaling model, the conditional probabilities for the manifest variables must be ordered so that one class can be interpreted, in some sense, as higher than, or more extreme than, the other group (e.g., $\pi_{11}^{\bar{A}X} > \pi_{12}^{\bar{A}X}$, $\pi_{11}^{\bar{B}X} > \pi_{12}^{\bar{B}X}$, etc.). Extreme-types models are interpretable most easily when the estimated conditional probabilities in one of the classes are all relatively large (.8 or higher, say) while those in the other latent class are all relatively small (.2 or smaller, say). In practice, the definition of "extreme" may be relaxed, especially when the variables are related to behaviors that are relatively infrequent for all respondents. Programs such as MLLSA, LCAG, and LEM do not currently allow specification of the explicit ordering relations given previously. However, as discussed in Section 3.B, in the context of the cheating data example, this is often not a problem in practice. For V dichotomous variables, the observed data comprise the frequencies for the 2^V response vectors. When $V = 3$, and $2^3 = 8$, an unconstrained two-class model always fits the data perfectly with no degrees of freedom for assessing model fit because there is one latent class proportion and six conditional probabilities that must be estimated (note that parameters for an unconstrained model with only two manifest variables cannot be estimated). Models that provide perfect fit to the data are referred to as saturated (or just identified). Thus, in practical applications it is desirable to have four or more manifest variables for the purposes of fitting extreme-types models.

Application to Medical Diagnosis

Even though saturated models always provide perfect fit to the data, it can often be important to estimate and interpret parameters for such models. An interesting example in the area of medical diagnosis was summarized by Walter and Irwig (1988). The frequencies summarized in Table 3.1 are for three medical personnel, labeled A, B, and C, each of whom is experienced in reading X rays, who were presented with chest X rays for 1692 males employed in asbestos mines and mills. The diagnostic categories were the presence (i.e., 1) or

TABLE 3.1
Frequencies for Pleural
Thickening Data

X-Ray Reader {ABC}	Frequency
{000}	1513
{100}	21
{010}	59
{110}	11
{001}	23
{101}	19
{011}	12
{111}	34
Total	1692

absence (i.e., 0) of thickening of the lung tissues (so-called pleural thickening). Prior to considering latent class analysis, we note that a model of complete independence provided poor fit for these data (G^2 value of 429.136 with 4 degrees of freedom, $p < .001$).

For the special case of three variables and two latent classes, an unconstrained latent class model for a response vector, y_s, can be written

$$P(y_s) = \pi_1^X \times \pi_{i1}^{\bar{A}X} \times \pi_{j1}^{\bar{B}X} \times \pi_{k1}^{\bar{C}X} + \pi_2^X \times \pi_{i2}^{\bar{A}X} \times \pi_{j2}^{\bar{B}X} \times \pi_{k2}^{\bar{C}X} \quad (3.1)$$

Note that in this example the variables, A, B, and C, are associated with the three X-ray readers and that the latent classes represent, in theory, disease groups (i.e., males for whom the X rays suggest or do not suggest pleural thickening). For these cases, we do not have the true diagnosis, or "gold standard," for purposes of comparison. Thus, the conditional probabilities in the model can be interpreted only loosely as rates of "correct" and "incorrect" diagnoses.

Clogg's program, MLLSA, was used to estimate the parameters for a two-class model, and results are shown as model I in Table 3.2. Note that the first latent class, comprising about 5% of the X rays, has consistently higher estimated probabilities for positive readings and, thus, can be interpreted as representing the diseased group. For two-class models, a useful comparative statistic for each variable (in this case, X-ray reader) is the odds ratio in favor of the first latent class. For the first reader, for example, the odds ratio is

$$\frac{\hat{\pi}_{11}^{\bar{A}X}/(1 - \hat{\pi}_{11}^{\bar{A}X})}{(1 - \hat{\pi}_{12}^{\bar{A}X})/\hat{\pi}_{12}^{\bar{A}X}} = \frac{\hat{\pi}_{11}^{\bar{A}X} \times (1 - \hat{\pi}_{12}^{\bar{A}X})}{(1 - \hat{\pi}_{11}^{\bar{A}X}) \times \hat{\pi}_{12}^{\bar{A}X}}$$
$$= \frac{.749 \times (1 - .010)}{(1 - .749) \times .010} = 295.42.$$

For readers B and C, the odds ratios are, respectively, 49.66 and 292.68. These large odds ratios suggest that the two classes, diseased and nondiseased, are highly discriminable.

The additive inverses of the conditional probabilities for the first latent class represent estimated rates of "false negatives" (e.g., for reader A, the false negative rate is $1 - \hat{\pi}_{11}^{\bar{A}X} = \hat{\pi}_{01}^{\bar{A}X} = 1 - .749 = .251$). We note that reader B appears to be somewhat inconsistent with the other two readers. The conditional probabilities for the second latent class ($\hat{\pi}_{12}^{\bar{A}X}$, etc.) represent estimated rates of "false positives" for the

TABLE 3.2
Models Fitted to Pleural Thickening Data

	Model	LC1	LC2	CP1	CP2	G^2	DF	Prob.	AIC*	BIC*	I_n
I	Unconstrained	0.055	0.945	.749, .643, .765	.010, .035, .011	0.000	0	n/a	0	0	0
II	Homogeneous	0.055	0.945	.716, .716, .716	.019, .019, .019	27.411	4	0.000	19.411	-2.324	0.018
III	Reader B, Het	0.055	0.946	.757, .643, .757	.010, .035, .010	0.134	2	0.935	-3.866	-14.733	0.001
IV	Reader B, FN	0.053	0.947	.737, .705, .737	.019, .019, .019	27.251	3	0.000	21.251	4.950	0.017
V	Reader B, FP	0.055	0.945	.713, .713, .713	.011, .034, .011	2.987	3	0.394	-3.013	-19.314	0.004

NOTES: Het denotes heterogeneous; FN denotes false negatives; FP denotes false positives; n/a denotes not applicable.

three X-ray readers. We note, again, that the second reader seems to be somewhat different than the other two readers; we will return to this point later.

Overall, the latent class model is consistent with the notion of two types of X rays representing men with and without plural thickening. Although the rates of false positives and false negatives appear reasonably consistent for readers A and C, reader B appears to be somewhat deviant. By considering various constrained latent class models, the consistency among the readers can be explored in more detail. We consider four additional models as summarized in Table 3.2. Note that the estimated proportions in the two latent classes are very similar across all five models. For model II, the rates of false positives are constrained to be equal for the readers and the rates of false negatives are similarly constrained. The resulting model posits homogeneity for the readers, where the specific restrictions are $\pi_{11}^{\bar{A}X} = \pi_{11}^{\bar{B}X} = \pi_{11}^{\bar{C}X}$ and $\pi_{12}^{\bar{A}X} = \pi_{12}^{\bar{B}X} = \pi_{12}^{\bar{C}X}$. This model, which has 4 degrees of freedom because the restrictions have, in effect, reduced six conditional probabilities to just two homogeneous values, fits the data very poorly. For model III, which provides good fit to the data, the homogeneity restrictions were relaxed to allow reader B to have unique conditional probabilities. The degrees of freedom for this model are 2 because the restrictions $\pi_{11}^{\bar{A}X} = \pi_{11}^{\bar{C}X}$ and $\pi_{12}^{\bar{A}X} = \pi_{12}^{\bar{C}X}$ have been imposed. For models IV and V, homogeneity is imposed in either the first or second latent class, respectively. The former of these models does not fit the data well, but the latter with 3 degrees of freedom, based on the restrictions $\pi_{11}^{\bar{A}X} = \pi_{11}^{\bar{B}X} = \pi_{11}^{\bar{C}X}$ and $\pi_{12}^{\bar{A}X} = \pi_{12}^{\bar{C}X}$, does fit well.

Based on G^2 values, both models III and V are reasonable choices for the pleural thickening data. Whereas these models are nested in the sense that model V is a restricted form of model III and whereas these restrictions involve equality of conditional probabilities, but do not involve restricting any parameter to a boundary value, the two models may be compared using a chi-square difference test. This difference is $2.987 - .134 = 2.854$ with $3 - 2 = 1$ degrees of freedom ($p = .108$). This value is nonsignificant at conventional levels and suggests that the simpler model V does not result in a significantly worse fit than model III. Note that this result is consistent with applying a min(BIC*) strategy, but is at variance with the application of a min(AIC*) that results in selection of model III. The estimated latent class proportions and the conditional probabilities for the first latent class are very similar for the two models. In an actual research context, the choice between them might be considered to be of little consequence or there may be substantive considerations that would tip the balance in favor of one of the two models.

B. Cheating Data Example

To introduce and illustrate some additional notions relative to model choice, we consider data from a survey of junior and senior undergraduate university students that contained ten dichotomous (yes/no) questions relating to academic cheating (Dayton and Scheers, 1997). Four of the questions concerned attitudes, rather than behaviors, and are not considered here. In addition, one question, which concerned knowledge of bribes for changing grades, was answered positively by only five students and was omitted from analysis. For the remaining five questions, students were asked whether or not, during their undergraduate years, they

Had lied to avoid taking an exam
Had lied to avoid handing a term paper in on time
Had purchased a term paper to hand in as their own
Had obtained a copy of an exam prior to taking the exam
Had copied answers during an exam from someone sitting near to them

Students from four colleges who had achieved at least junior standing were sampled randomly to receive the survey by mail. Based on a total of 319 responses, the proportions of students acknowledging engaging in the five behaviors were, respectively, .11, .12, .03, .04, and .21. The third and fourth questions, which represent relatively serious types of academic cheating, were responded to positively by only 10 and 13 students, respectively. Thus, for purposes of analysis, these two questions were combined into one question, labeled C, that was scored 1 if a "yes" response was given to either the third or fourth cheating item. The proportion of positive responses for item C was .07. In the following analyses, the first two cheating items are labeled A and B, respectively, while the fifth item is labeled D (frequencies for the response vectors are shown in Table 3.4).

Before turning to other issues, there is the question of whether or not it is worthwhile to fit a latent class model to the cheating data. A preliminary test is provided by assuming a model of complete independence for responses to the questions using the cited proportions.

This model yields a G^2 statistic equal to 62.586, which, with 11 degrees of freedom, represents poor fit ($p < .001$) and suggests that more complex modeling is justified.

From previous research, there is support for the notion that there is a subgroup of students who, relatively speaking, are persistent cheaters. This subgroup is distinct from other students who are referred to as noncheaters, although they, too, may cheat on an opportunistic basis, but show such behavior at a relatively lower rate than do persistent cheaters. Given this conceptualization, we consider fitting a two-class latent class model of the form

$$P(y_s) = \pi_1^X \times \pi_{i1}^{\bar{A}X} \times \pi_{j1}^{\bar{B}X} \times \pi_{k1}^{\bar{C}X} \times \pi_{l1}^{\bar{D}X}$$
$$+ \pi_2^X \times \pi_{i2}^{\bar{A}X} \times \pi_{j2}^{\bar{B}X} \times \pi_{k2}^{\bar{C}X} \times \pi_{l2}^{\bar{D}X} \tag{3.2}$$

where the responses $i = j = k = l = 1$ represent "yes" responses to the cheating questions and the responses $i = j = k = l = 0$ represent "no" responses. If we assume that the first class represents persistent cheaters, then the expectation is that, for each survey question, members of this class are more likely to answer "yes" to the question. Thus, π_1^X represents the proportion of persistent cheaters if we assume that the conditional probabilities for the questions obey the inequalities $\pi_{11}^{\bar{A}X} > \pi_{12}^{\bar{A}X}$, $\pi_{11}^{\bar{B}X} > \pi_{12}^{\bar{B}X}$, $\pi_{11}^{\bar{C}X} > \pi_{12}^{\bar{C}X}$, and $\pi_{11}^{\bar{D}X} > \pi_{12}^{\bar{D}X}$. As a matter of convenience, our analytic strategy is to fit the model without considering the inequality restrictions, because latent class programs such as MLLSA, LCAG, and LEM do not currently include provisions for imposing inequality restrictions.[4]

Parameter estimates for the two-class model, obtained from Clogg's MLLSA program, are summarized in Table 3.3. The column labeled "CP" contains estimates for the conditional probabilities with the latent class proportion at the bottom and the columns labeled "SE(J)," "SE(B)," and "SE(N)" contains three different estimates for standard errors that will be explained subsequently. The first latent class, estimated to contain about 16% of respondents, appears to be consistent with the notion of persistent cheaters, because all conditional probabilities that correspond to "yes" responses in this class are considerably larger than in the second class. Also, the odds ratios in favor of the first latent class range from 2.720 for item D to 79.525 for item A. The G^2 goodness-of-fit statistic is 7.764 with degrees of freedom $16 - 9 - 1 = 6$, and the model fits the data in an acceptable fashion ($p = .256$). The Pearson statistic is only slightly larger

TABLE 3.3
Two-Class Solution for Cheating Data

Item	Questionnaire Item	Class 1				Class 2				Odds Ratio
		CP	SE(J)	SE(B)	SE(N)	CP	SE(J)	SE(B)	SE(N)	
A	Have lied to avoid taking an exam	0.579	0.198	0.170	0.181	0.017	0.030	0.019	0.030	79.525
B	Have lied to avoid handing a term paper in on time	0.591	0.189	0.155	0.176	0.030	0.033	0.024	0.032	46.721
C	Have purchased a term paper to hand in as their own OR have obtained a copy of an exam prior to taking the exam	0.217	0.091	0.080	0.085	0.037	0.016	0.014	0.014	7.213
D	Have copied answers during an exam from someone sitting nearby	0.377	0.113	0.097	0.097	0.182	0.027	0.025	0.026	2.720
	Latent class proportion	0.160	0.084	0.052	0.077	0.840	0.084	0.052	0.077	

$(X^2 = 8.329, p = .215)$ and the Read and Cressie I^2 statistics is intermediate in value $(I^2 = 8.099, p = .231)$. The descriptive index of dissimilarity, $I_D = .032$, computed directly from the entries in Table 3.4, also supports the notion that a two-class model is satisfactory for these cheating data.

Although the two-class model provided a satisfactory fit, it is possible that a more complex model would provide an even better fit. Therefore, an unrestricted three-class model was fitted to the cheating data: The resulting G^2 value, .181 ($p = .913$), also suggests a satisfactory fit. There is a complication with the three-class model in that the usual computation for degrees of freedom would be $16 - 14 - 1 = 1$, but investigation of the rank of the asymptotic covariance matrix (as reported by MLLSA) shows that there is some redundancy among the parameters and that the correct degrees of freedom for an identified model are 2 (note that this is a general result for three latent classes fitted to four dichotomous response variables). Although the two-class and three-class models are nested, they cannot, for the technical reasons cited in Chapter 2, be compared using the difference in G^2 values. However, the AIC* statistics are, respectively,

TABLE 3.4
Two Class Model Fitted to Academic Cheating Data

{ABCD}	Freq.	Exp. Freq.	X^2	Per(X^2)
{0000}	207	205.71	0.008	0.001
{1000}	10	9.35	0.045	0.005
{0100}	13	12.31	0.039	0.005
{1100}	11	8.60	0.670	0.080
{0010}	7	8.96	0.429	0.051
{1010}	1	1.76	0.328	0.039
{0110}	1	1.95	0.463	0.056
{1110}	1	2.35	0.776	0.093
{0001}	46	47.42	0.043	0.005
{1001}	3	4.33	0.409	0.049
{0101}	4	5.11	0.241	0.029
{1101}	4	5.17	0.265	0.032
{0011}	5	2.45	2.654	0.318
{1011}	2	1.01	0.970	0.116
{0111}	2	1.09	0.760	0.091
{1111}	2	1.42	0.237	0.028
Total	319	318.99	8.335	

7.764 − 2(6) = −4.236 and .181 − 2(2) = −3.819, and, thus, a min(AIC*) strategy results in the selection of the two-class model as preferable as does a min(BIC*) strategy (corresponding values of −26.827 and −11.349).

Expected frequencies for the two-class model are summarized in Table 3.4, where we note that, in general, the cells with large observed frequencies of respondents are fitted very well by the model. To aid in interpreting the discrepancies that do exist, two additional columns are presented. The entries in the "X^2" column are components, $(n_s - \hat{n}_s)^2/\hat{n}_s$, of the Pearson chi-square statistic and the entries in the "Per(X^2)" column are these components expressed as a percentage of the overall Pearson chi-square (note that the column sum for X^2 is 8.335 rather than the value of 8.329 reported previously; this difference is due to rounding errors that resulted from displaying expected frequencies with only two decimal places in the MLLSA output). Again, we note the excellent agreement for cells with larger frequencies and, also, that nearly a third of the Pearson chi-square value results from underestimating the frequency for those who answered "yes" to both questions C and D. Whereas our model fits the data well, these diagnostics are not of any additional help, but in other cases, examination of these components may be useful to suggest better fitting models.

C. Estimating the π^* Measure of Fit

Whereas the foregoing latent class estimates and statistics can be computed directly or indirectly using available computer programs such as MLLSA, calculation of the Rudas, Clogg, and Lindsay (1994) index of model fit, π^*, requires special programming (note that the program, MIXIT, developed by Rudas and Clogg computes π^* for conventional two-way contingency tables, but not for mixture models such as those represented by latent class analysis). Our approach to computing π^* is based on a generalization of the nonlinear programming algorithm presented by Xi (1994) and can be implemented in a spreadsheet program such as Microsoft Excel. The computational procedure is described in detail in the Web site, //www.inform.umd.edu/EDUC/Depts/EDMS. This procedure involves the creation of a special set of expected frequencies that are denoted $\tilde{n}_s$. For the academic cheating data, the results of Excel

TABLE 3.5

Two-Class π^* Model Fitted to Academic
Cheating Data

Item {ABCD}	Freq.	E1	E2	Expected
{0000}	207	202.999	4.001	207.000
{1000}	10	4.469	5.531	10.000
{0100}	13	5.124	7.876	13.000
{1100}	11	0.113	10.887	11.000
{0010}	7	6.634	0.366	7.000
{1010}	1	0.146	0.506	0.652
{0110}	1	0.167	0.721	0.888
{1110}	1	0.004	0.996	1.000
{0001}	46	44.539	1.461	46.000
{1001}	3	0.981	2.019	3.000
{0101}	4	1.124	2.876	4.000
{1101}	4	0.025	3.975	4.000
{0011}	5	1.456	0.134	1.589
{1011}	2	0.032	0.185	0.217
{0111}	2	0.037	0.263	0.300
{1111}	2	0.001	0.364	0.365
Total	319	267.850	42.161	310.011 $= S$
				0.028 $= \pi^\circ$

spreadsheet computations for π^* are shown in Table 3.5. The column labeled "Expected" contains the $\tilde{n}_s$ values, the columns labeled $E1$ and $E2$ are intermediate computations explained in the Web site, and $S = \sum_{s=1}^{2^4} \tilde{n}_s = 310.01$ is maximized, which results in the value $\pi^* = 1 - 310.01/319 = .028$. The interpretation of π^* is that the frequencies for the response vectors represent a two-class model perfectly if the appropriate 2.8% of the cases are removed. As a rule of thumb, values of π^* less than about .10 represent reasonable model fit. Note that the expected frequencies, $\tilde{n}_s$, are not integer values, but if integer values are desired, nonlinear optimization procedures ordinarily allow integer restrictions on the solution, although convergence to integer values may be very difficult to achieve in practice.

D. Estimating Standard Errors

For the cheating data, the estimated proportion of persistent cheaters, $\hat{\pi}_1^X$, is equal to .160. However, latent class programs such as

MLLSA or LCAG do not provide estimates of standard errors that can be used to construct confidence intervals for this estimate or for any of the conditional probabilities for the various cheating questions. To remedy this deficiency, jackknife and/or bootstrap estimates of standard errors can be found; both are summarized in Table 3.3 [i.e., the columns labeled "SE(J)" and "SE(B)," respectively]. To conduct the jackknife, a total of 16 MLLSA analyses, as described in the previous chapter, were performed for the cheating data. The spreadsheet computations for the standard error for the estimated latent class proportion, $\hat{\pi}_1^X = .160$, are summarized in Table 3.6. The column labeled "ThJ" contains the estimates from a set of 16 MLLSA analyses [e.g., the entry for the response vector {0000}, .1599, results when the frequency for that vector is reduced by 1 (to 206)] and the remaining columns are for components of the computations in Equation 2.11. In addition, using the Gauss program presented in the Web site, (//www.inform.umd.edu/EDUC/Depts/EDMS), 500 bootstrap samples were generated based on the parameter estimates from Table 3.3. The summary output from the Gauss program is presented in Table 3.7.

TABLE 3.6
Two-Class Model Fitted to Academic Cheating Data

Item {ABCD}	Freq.	ThJ	F*ThJ	F*Dev²
{0000}	207	0.160	33.110	0.0000
{1000}	10	0.163	1.629	0.0001
{0100}	13	0.163	2.117	0.0001
{1100}	11	0.150	1.646	0.0011
{0010}	7	0.164	1.147	0.0001
{1010}	1	0.140	0.140	0.0004
{0110}	1	0.140	0.140	0.0004
{1110}	1	0.169	0.169	0.0001
{0001}	46	0.161	7.392	0.0000
{1001}	3	0.149	0.446	0.0004
{0101}	4	0.150	0.599	0.0004
{1101}	4	0.165	0.661	0.0001
{0011}	5	0.156	0.779	0.0001
{1011}	2	0.144	0.289	0.0005
{0111}	2	0.143	0.286	0.0006
{1111}	2	0.196	0.392	0.0027
Total	319		0.160	0.0071 = VAR(J)
				0.0842 = SE(J)

TABLE 3.7
Parametric Bootstrap Standard Error
Estimates for Cheating Data

(gauss) run c: \ gauss\ sim\ sim1.gss
step   1.0000000
step   2.0000000
...
step   499.00000
step   500.00000

   1.0000000=hours
   4.0000000=minutes
   4.0000000=second
   18.000000=hundredth/second

Did not coverge   24.000000 times
Mean LC proportions:
  0.152481   0.847519
Mean conditional probabilities:
  LC1: 0.631627   0.631301   0.241568   0.393185
  LC2: 0.019135   0.031992   0.036917   0.184064
Standard errors for LC proportions:
  0.051890943   0.051890943
Standard errors for conditional probabilities:
  LC1: 0.169996   0.154758   0.080174   0.096863
  LC2: 0.019088   0.023725   0.014110   0.025112

Note that the standard error estimates from the jackknife are larger than those from the bootstrap, although the difference is substantial only for $\hat{\pi}_1^X$, where the estimated standard error is .084 for the jackknife, but only .052 for the parametric bootstrap. There is some evidence that jackknife estimates tend to be somewhat too large, although for relatively large samples, on average, there does not seem to be much difference in the precision of jackknife and bootstrap standard error estimates in the context of latent class analysis (Bolesta, 1998). For the purposes of comparison, a third set of standard error estimates, labeled "SE(N)," is included in Table 3.3. These estimates resulted from the program LEM that uses a Newton estimation procedure. Note that, for this example, the Newton estimates tend to be between those from the jackknife and parametric bootstrap procedure, but are not consistently closer to either set of estimates.

The use of the bootstrap standard error estimate for the latent class proportion results in, for example, a 90% confidence interval around $\hat{\pi}_1^X = .16$ of $\{.07, .25\}$. This relatively wide interval suggests that the proportion of persistent cheaters is not estimated with great precision based on our sample of 319 students. There is, in general, greater sampling variability associated with latent class proportions than for similar quantities that are estimated directly by simple random sampling (SRS). For example, if a proportion were estimated to be .160 based on an SRS of 319 cases, the standard error would be $\sqrt{(.160 \times .840)/318} = .021$. Note that the parametric bootstrap estimate, .052, is larger by a factor of over 2.5.

E. Classifying Respondents

Using Bayes' theorem, Equation 2.3, the posterior probability for each response vector and latent class can be computed and, based on these, cases can be classified as representing or not representing the first latent class (i.e., persistent cheaters). For each response vector, the MLLSA program displays the largest (in this case, larger) posterior probability, $\max[P(t \mid \mathbf{y}_s)]$, in a column labeled "Modal P=" and shows the Bayes classification in a column labeled "Class=." However, to show how these results are attained, Table 3.8 displays these computations, as well as the posterior odds, for membership in the persistent cheater latent class. Note that all response vectors, except $\{0000\}$, $\{0010\}$, $\{0001\}$, and $\{0011\}$, result in classification as persistent cheaters. Thus, noncheaters are characterized by responding "no" to both of the first two questions. If one considers a score based on the number of cheating questions to which "yes" answers are given, this is ambiguous for persons scoring 1 or 2, because these response vectors are not consistently classified (i.e., the vector $\{1000\}$ is classified as a persistent cheater, whereas the vector $\{0010\}$ is not).

Various descriptive measures can be derived from the classification results. Clogg's program, MLLSA, reports a "Percent Correctly Allocated" that represents a summation of the components of the form $P(\mathbf{y}_s \mid t) \times \hat{\pi}_t^X$, but only those components that are consistent with the classification decision for the various response vectors. That is, the summation contains the elements $P(\mathbf{y}_s \mid t = 1) \times \hat{\pi}_1^X$ when $\mathbf{y}_s$ is classified in the first latent class, but contains the elements $P(\mathbf{y}_s \mid t = 2) \times \hat{\pi}_2^X$ when $\mathbf{y}_s$ is classified in the second latent class (note that this is an alternative, but equivalent, approach to

TABLE 3.8

Bayes Classification for Academic Cheating Data

Item {ABCD}	Freq.	$P(y \mid t = 1)^* \pi_1^y$	$P(y \mid t = 2)^* \pi_2^y$	Predicted Class	$P(t = 1 \mid y)$	Odds
{0000}	207	0.013	0.631	2	0.021	0.021
{1000}	10	0.018	0.011	1	0.629	1.698
{0100}	13	0.019	0.019	1	0.502	1.010
{1100}	11	0.027	0.000	1	0.988	80.704
{0010}	7	0.004	0.024	2	0.132	0.152
{1010}	1	0.005	0.000	1	0.924	12.161
{0110}	1	0.005	0.001	1	0.879	7.232
{1110}	1	0.007	0.000	1	0.998	577.856
{0001}	46	0.008	0.141	2	0.055	0.058
{1001}	3	0.011	0.002	1	0.822	4.621
{0101}	4	0.012	0.004	1	0.733	2.748
{1101}	4	0.016	0.000	1	0.995	219.592
{0011}	5	0.002	0.005	2	0.293	0.414
{1011}	2	0.003	0.000	1	0.971	33.089
{0111}	2	0.003	0.000	1	0.952	19.677
{1111}	2	0.004	0.000	1	0.999	1572.317
Total	319	0.160	0.840			

computing the quantity, P_c, presented as Equation 2.4 in Chapter 2). For the cheating data, this statistic equals 93.4% and can be interpreted as representing an estimate of the proportion of correct classifications that would be made for the population as a whole. However, this large success rate is moderated by the fact that the second latent class is estimated to contain about 84% of the population and, thus, a success rate of 84% can be attained by simply classifying everyone as noncheaters (i.e., in latent class 2). To correct for this, the lambda statistic is recommended. For the cheating data, $\lambda = (.934 - .840)/(1 - .840) = .588$, which can be interpreted as meaning that using Bayes' theorem for the purposes of classifications results in a 59% improvement over chance assignment based only on the larger latent class proportion.

F. Validation

Often, there is additional information available for respondents that can be used to assess the construct validity of the latent class structure. In the case of the cheating data, self-reported grade point aver-

TABLE 3.9
Validation for Academic Cheating Data

GPA	Frequency		Total	Proportion		Odds
	Class 1	Class 2		Class 1	Class 2	
1	26	74	100	0.481	0.284	2.846
2	18	86	104	0.333	0.330	4.778
3	6	42	48	0.111	0.161	7.000
4	2	32	34	0.037	0.123	16.000
5	2	27	29	0.037	0.103	13.500
Total	54	261	315	0.999	1.001	4.833

age (GPA) in a set of five ordered categories (where 5 was the highest GPA) was available for 315 of the 319 students, and it is reasonable to expect that persistent cheaters would be represented more heavily in the lower GPA categories. One approach to investigating this relation is to cross-tabulate the GPA categories and the Bayes classifications from Table 3.8, and this is shown in Table 3.9. The column labeled "Odds" presents, for each GPA category, the odds in favor of the noncheater class, and we note that these odds increase dramatically when moving from the lower to the higher GPA categories. Viewed somewhat differently, about 50% of the students identified as persistent cheaters (i.e., 26 out of 54) are in the lowest GPA category and a total of only 7.4% are in the highest two GPA categories, whereas these percentages for noncheaters are 28.4% and 22.6%, respectively. A more sophisticated approach to the validation issue involves modeling the relationship between the proportion of cases in each latent class and one or more additional variable such as GPA. This concomitant variable approach is presented and illustrated for the academic cheating data in Chapter 7.

G. Mixture-Binomial Model

In the extreme-types model, each variable has a unique conditional probability for a positive response for each latent class, and the frequencies for the various response vectors represent the statistics required to fit this model to data. However, if the number of variables is fairly large and the sample size is limited, there will be insufficient information to estimate parameters in the extreme-types model. Thus, it is of interest to consider restricted models that do not require the frequencies for the response vectors for the purposes of estimating parameters. A relatively simple type of model is the mixture of binomial processes. This model can be derived as a restricted version of the extreme-types model defined in Equation 3.2. In essence, we assume that all variables within a given latent class have the same conditional probability of a positive response (and, hence, of a negative response). For four variables, the appropriate restrictions are $\pi_{11}^{\bar{A}X} = \pi_{11}^{\bar{B}X} = \pi_{11}^{\bar{C}X} = \pi_{11}^{\bar{D}X} \equiv \pi_{11}^{X}$ and $\pi_{12}^{\bar{A}X} = \pi_{12}^{\bar{B}X} = \pi_{12}^{\bar{C}X} = \pi_{12}^{\bar{D}X} \equiv \pi_{12}^{X}$, where π_{11}^{X} and π_{12}^{X} are the binomial rate parameters for the first and second latent classes, respectively. Note that for this model, no matter how many variables are involved, there are only three parameters to estimate—the latent class proportion, π_{1}^{X}, and the two conditional probabilities, π_{11}^{X} and π_{12}^{X}. For this simple model, which we call the mixture-binomial model, the only information required to estimate the parameters is the set of frequencies for the scores that represent the number of 1 responses for the variables. For example, with five manifest variables, the data required are the frequencies for the scores 0, 1, 2, 3, 4, and 5. For most data situations, it is unrealistic to assume that variables have the same response rate for positive responses. However, we can adopt the view that the binomial parameters, π_{11}^{X} and π_{12}^{X}, represent **average** rates for the variables within the latent classes.

We consider a simple set of data where it is not feasible to fit complex models because of the limited sample size. In Table 3.10, the column labeled "Score" represents the numbers of correct responses, with corresponding frequencies and proportions, on a quiz given to students in a multivariate statistics course I teach each Spring. It has been my practice to present basic elements of matrix algebra early in the course and to test on this material using a ten-item quiz. Although the specific quiz items vary from year to year, the content being sampled is essentially the same over time. The frequencies in the table are for a total of 247 students over a period of ten years (it is somewhat reassuring to note that no student in this period of time attained a score of 0). If X_s represents a quiz score and F_s represents the corresponding frequency, then the average **proportion** of items answered correctly is $(\sum_{s=0}^{10} F_s \times X_s)/2470 = .713$. Based on this proportion, a single binomial distribution, corresponding to a one-class "latent class" model, can be fitted to the data. For a score of X, the expected binomial proportion is given by

42

TABLE 3.10
Binomial Distributions Fitted to Quiz Data

Score	Frequency	Observed Proportion	Expected Proportion One Binomial	Two Binomials
0	0	0.000	0.000	0.000
1	2	0.008	0.000	0.002
2	4	0.016	0.001	0.011
3	5	0.020	0.007	0.034
4	17	0.069	0.030	0.071
5	24	0.097	0.090	0.104
6	37	0.150	0.187	0.126
7	32	0.130	0.266	0.155
8	60	0.243	0.248	0.201
9	37	0.150	0.137	0.199
10	29	0.117	0.034	0.096
Total	247	1.000	1.000	0.999

$_{10}C_X \times (.713)^X \times (1-.713)^{10-X}$, where $_{10}C_X$ is the combinations of ten things taken X at a time [i.e., $10!/X! \times (10-X)!$, where the factorial $X! = X \times (X-1) \times (X-2) \times \cdots \times 1$]. These expected binomial proportions, labeled "One Binomial" in Table 3.10, are displayed, along with the observed proportion for each score, in the top panel of Figure 3.1. Note that the binomial distribution appears to provide a poor fit to the data and, in fact, the G^2 goodness-of-fit value is 142.69 with 9 degrees of freedom ($p < .001$).

The quiz data in Figure 3.1 appear to be distinctly bimodal with relative maxima at scores of 6 and 8. This suggests that a mixture model could be a better representation for the quiz score distribution. Estimating parameters for the mixture-binomial model using programs such as MLLSA or LCAG is problematic because these programs require that the data be inputted in terms of response vectors and, in the present case, there are $2^{10} = 1,024$ potential response vectors, most of which do not, in fact, occur, because the sample size is only 247. For this reason, we turn to special programming for parameter estimation provided by the program MODEL3 (Dayton and Macready, 1977), although it is fairly straightforward to use nonlinear programming techniques such as those described in Section 3.C for the mixture-binomial model. The estimates and associated standard errors (in parentheses) for the quiz data are $\hat{\pi}_1^X = .600$ (.071),

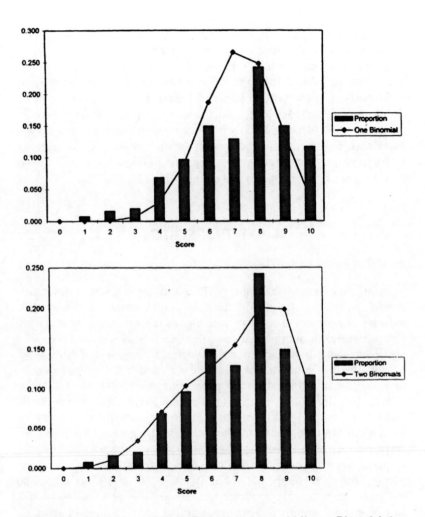

Figure 3.1. Comparison of Single Binomial (top) and Mixture Binomial (bottom) for Quiz Data

$\hat{\pi}_{11}^{X} = .832$ (.018), and $\hat{\pi}_{12}^{X} = .536$ (.031). (Note: Using the jackknife to estimate the standard errors results in the somewhat larger values .087, .022, and .038, respectively.) These results can be interpreted as suggesting that there is a relatively high performance group comprising 60% of students that, on average, attain a score of about 83% (i.e.,

8.3 items correct) and a second, relatively lower, performing group comprising 40% of students that, on average, attain a score of about 54% (i.e., 5.4 items correct). Expected frequencies using this model are shown in Table 3.10 in the column labeled "Two Binomials" and are displayed in the bottom panel of Figure 3.1. The improved fit to the data of this model is apparent from the figure and from the G^2 goodness-of-fit value of 13.37 with 7 degrees of freedom ($p = .064$). Note that, in general, for the case of two latent classes, the degrees of freedom for assessing the fit of the mixture-binomial model to the $V + 1$ scores for V variables are equal to $V - 3$.

4. LINEAR SCALES

A. Introduction

If, in some sense, successful performance on a given variable depends upon successful performance on one or more other variables, then we can consider the set of variables as representative of a hierarchic structure. In the simplest case, the dependencies constitute a single, linear scale. In the area of psychological scaling, sets of attitude or personality test items that show a linear structure often are referred to as Guttman scales after the pioneering work of Louis Guttman (1947), but the notion can be applied equally well to achievement tasks or behavioral tasks. In discussing linear scales, it is convenient to think of the theoretical structure as comprising a set of permissible response vectors. For dichotomous response variables, the permissible response vectors comprise 0s and 1s in a systematic pattern. For example, with three mathematics test items, A, B, and C, assume that the items form a scale in the sense that knowledge required to answer item A correctly is also necessary to correctly answer item B, and that knowledge required to answer both items A and B correctly is necessary to correctly answer item C. This set of prerequisite relations can be represented as $A \rightarrow B \rightarrow C$, where the right arrow ($\rightarrow$) is read as "is prerequisite to." In terms of observed response vectors, only certain patterns are consistent with the hypothesized linear, or Guttman, scale. In particular, the response vectors {000}, {100}, {110}, and {111} are permissible, assuming that 1 indicates a correct response and 0 indicates an incorrect response to an item. Note that the response vector {000} is included as a permissible response vector because some respondents may answer all three items incorrectly. Other possible response vectors, such as {010}, for example, are not permissible, because the scale is based on the assumption that knowledge of item A is required for a correct answer to item B. Of course, in practice, even if the linear scale correctly characterizes the population of respondents, response vectors other than the permissible vectors are likely to be observed due to respondent errors such as inadvertently selecting an incorrect response, guessing the correct response, etc.

Although we illustrated a linear scale using achievement test items, similar reasoning can be applied to items that are subjective in nature. For example, three attitude items might be constructed so that they are intended to represent increasingly stronger beliefs concerning some social issue. Thus, endorsement of or agreement with item B implies endorsement of item A, and endorsement of item C implies endorsement of both items A and B. In the remainder of this chapter, we present some fundamental models for linear scales, and in the next chapter, we consider structures that are more complex than linear.

B. Models with Response Errors

Proctor Model

Consider three dichotomous variables, A, B, and C, that, in theory, obey the following two prerequisite relations: a 1 response (e.g., "yes" or "correct") on variable B requires a 1 response on variable A and a 1 response on variable C requires 1 responses on both variables A and B. Then the permissible response vectors, {000}, {100}, {110}, and {111}, represent a linear scale. Proctor (1970) proposed a model in which it is assumed that the permissible response vectors represent "true types" in the population, but that there is some constant probability of error for all of the variables.

The Proctor model can be formulated as a latent class model with a separate latent class corresponding to each of the true types and with appropriate restrictions on the conditional probabilities for the variables. In particular, the proportion of each true type in the population is the latent class proportion, π_t^X, where $t = 1$ corresponds to the permissible response vector {000}, $t = 2$ corresponds to the permissible response vector {100}, and so forth. In general, for linear scales, the number of latent classes is one more than the number of response variables (i.e., $V + 1$). The specific restrictions on the conditional probabilities vary across the latent classes and are

$$\pi_{11}^{\bar{A}X} = \pi_{11}^{\bar{B}X} = \pi_{11}^{\bar{C}X} = \pi_{12}^{\bar{B}X} = \pi_{12}^{\bar{C}X} = \pi_{13}^{\bar{C}X} = \cdots$$

$$= \pi_{02}^{\bar{A}X} = \pi_{03}^{\bar{A}X} = \pi_{03}^{\bar{B}X} = \pi_{04}^{\bar{A}X} = \pi_{04}^{\bar{B}X} = \pi_{04}^{\bar{C}X} \equiv \pi_e \tag{4.1}$$

To simplify notation, the constrained probability in Equation 4.1 has been represented by π_e. Then, for any respondent in a given latent class, or true type, the probability of a response that is inconsistent with the true type (i.e., an error) is π_e, whereas the probability of a consistent response (i.e., a nonerror) is $1 - \pi_e$. Consider, for example, a respondent in the latent class $t = 2$ corresponding to the permissible response vector $\{100\}$. The probabilities associated with the various possible response vectors are

$$P(\mathbf{y}_s = \{000\} \mid t = 2) = \pi_e \times (1 - \pi_e)^2$$

$$P(\mathbf{y}_s = \{100\} \mid t = 2) = (1 - \pi_e)^3$$

$$P(\mathbf{y}_s = \{010\} \mid t = 2) = \pi_e^2 \times (1 - \pi_e)$$

$$P(\mathbf{y}_s = \{110\} \mid t = 2) = \pi_e \times (1 - \pi_e)^2$$

$$P(\mathbf{y}_s = \{001\} \mid t = 2) = \pi_e^2 \times (1 - \pi_e)$$

$$P(\mathbf{y}_s = \{101\} \mid t = 2) = \pi_e \times (1 - \pi_e)^2$$

$$P(\mathbf{y}_s = \{011\} \mid t = 2) = \pi_e^3$$

$$P(\mathbf{y}_s = \{111\} \mid t = 2) = \pi_e^2 \times (1 - \pi_e)$$

Note that the exponent for the term π_e is the number of discrepancies, or errors, when each response vector is compared with the permissible response, $\{100\}$, and that the exponent for the term $1 - \pi_e$ is the number of matches, or nonerrors, in the same comparison.

Given the restrictions in Equation 4.1, the Proctor model is a constrained form of the general latent class model presented in Equations 2.1 and 2.2 with four latent classes. The analysis follows the procedures developed in Chapter 2 and illustrated in Chapter 3; with the only special consideration is determination of the appropriate degrees of freedom for chi-square goodness-of-fit tests. The general rule for degrees of freedom, assuming that a model is identified, is to reduce the number of response vectors, 2^V, by the number of independent restrictions placed on the expected frequencies. The requirement that the sum of the expected frequencies be equal to the observed sample size always represents a restriction (i.e., $\sum_{s=1}^{8} \hat{n}_s = N$), and each independent parameter that is estimated from the observed frequencies represents an additional restriction. In the present case, the constant error rate, π_e, as well as three of the four latent class proportions, must be estimated from the data (the fourth latent class proportion is mathematically determined by the requirement that the sum of the latent class proportions must equal 1.0). Thus, the degrees of freedom for the three-item example are $8 - 5 = 3$. In general, for a Proctor model based on V variables, the degrees of freedom are $2^V - V - 2$.

Intrusion–Omission Error Model

Positing a single error rate, π_e, as in the Proctor model, is an oversimplification in many applications, but the model can be generalized in a number of ways. Dayton and Macready (1976) introduced the notion of two distinct types of errors: intrusion errors and omission errors. An intrusion error, with probability π_I, occurs if a 1 response (e.g., "yes" or "correct") is observed where the permissible response vector calls for a 0 response (e.g., "no" or "incorrect"). An omission error, with probability π_O, occurs if a 0 response is observed where the permissible response vector calls for a 1 response. The new set of restrictions that replaces Equation 4.1 is

$$
\begin{aligned}
\pi_{11}^{\bar{A}X} &= \pi_{11}^{\bar{B}X} = \pi_{11}^{\bar{C}X} = \pi_{12}^{\bar{B}X} = \pi_{12}^{\bar{C}X} = \pi_{13}^{\bar{C}X} \equiv \pi_I \\
\pi_{02}^{\bar{A}X} &= \pi_{03}^{\bar{A}X} = \pi_{03}^{\bar{B}X} = \pi_{04}^{\bar{A}X} = \pi_{04}^{\bar{B}X} = \pi_{04}^{\bar{C}X} \equiv \pi_O
\end{aligned}
\tag{4.2}
$$

With respect to the permissible response vector {100}, the probabilities associated with the various response vectors are

$$P(\mathbf{y}_s = \{000\} \mid t = 2) = \pi_O \times (1 - \pi_I)^2$$

$$P(\mathbf{y}_s = \{100\} \mid t = 2) = (1 - \pi_O) \times (1 - \pi_I)^2$$

$$P(\mathbf{y}_s = \{010\} \mid t = 2) = \pi_O \times \pi_I \times (1 - \pi_I)$$

$$P(\mathbf{y}_s = \{110\} \mid t = 2) = (1 - \pi_O) \times \pi_I \times (1 - \pi_I)$$

$$P(\mathbf{y}_s = \{001\} \mid t = 2) = \pi_O \times (1 - \pi_I) \times \pi_I$$

$$P(\mathbf{y}_s = \{101\} \mid t = 2) = (1 - \pi_O) \times (1 - \pi_I) \times \pi_I$$

$$P(\mathbf{y}_s = \{011\} \mid t = 2) = \pi_O \times \pi_I^2$$

$$P(\mathbf{y}_s = \{111\} \mid t = 2) = (1 - \pi_O) \times \pi_I^2$$

The analysis proceeds as for any latent class model except that the degrees of freedom for this specific case would be $8 - 6 = 2$ and, in general, are $2^V - V - 3$.

Other Error Models

In addition to the Proctor and intrusion–omission error models, it is possible to restrict conditional probabilities for the variables in other meaningful ways including variable-specific errors and latent-class-specific errors. In the first case, there are distinct error rates, π_{v1}, π_{v2}, and π_{v3}, for each of the V variables and the appropriate restrictions for the previous example are

$$\pi_{11}^{\bar{A}X} = \pi_{02}^{\bar{A}X} = \pi_{03}^{\bar{A}X} = \pi_{04}^{\bar{A}X} \equiv \pi_{v1}$$

$$\pi_{11}^{\bar{B}X} = \pi_{12}^{\bar{B}X} = \pi_{03}^{\bar{B}X} = \pi_{04}^{\bar{B}X} \equiv \pi_{v2} \qquad (4.3)$$

$$\pi_{11}^{\bar{C}X} = \pi_{12}^{\bar{C}X} = \pi_{13}^{\bar{C}X} = \pi_{04}^{\bar{C}X} \equiv \pi_{v3}$$

In the second case, there are distinct error rates, π_{c1}, π_{c2}, π_{c3}, and π_{c4}, for each of the T latent classes, and the appropriate restrictions for the previous example are

$$\pi_{11}^{\bar{A}X} = \pi_{11}^{\bar{B}X} = \pi_{11}^{\bar{C}X} \equiv \pi_{c1}$$

$$\pi_{02}^{\bar{A}X} = \pi_{12}^{\bar{B}X} = \pi_{12}^{\bar{C}X} \equiv \pi_{c2}$$

$$\pi_{03}^{\bar{A}X} = \pi_{03}^{\bar{B}X} = \pi_{13}^{\bar{C}X} \equiv \pi_{c3} \qquad (4.4)$$

$$\pi_{04}^{\bar{A}X} = \pi_{04}^{\bar{B}X} = \pi_{04}^{\bar{C}X} \equiv \pi_{c4}$$

In general, the degrees of freedom for assessing model fit for these two cases are, respectively, $2^V - 2V - 1$ and $2^V - V - T - 1 = 2^V - 2V - 2$. Note that with three variables, the former model has 1 degree of freedom and that the latter is saturated (i.e., has 0 degrees of freedom).

C. Clinical Scale Example

The mastery of left–right spatial tasks is regarded by pediatricians as an indication of normal development in children. Whitehouse, Dayton, and Eliot (1980) attempted to develop a simple developmental scale that could be used by clinicians. A set of 12 behavioral tasks was administered to 573 children between the ages of three and nine. These children had no known handicaps and were judged to be performing at "normal" academic levels for their age. The 12 tasks were grouped into three levels: (A) body identification (e.g., "Show me your LEFT hand"), (B) cross-midline identification (e.g., "Put your LEFT hand on your RIGHT knee"), and (C) reverse opposite identification (e.g., "Put your RIGHT hand on my LEFT knee"). After initial screening and analysis of the responses, a scoring scheme was devised using six of the twelve tasks, two from each of the three levels. The scoring for each level was such that performing both tasks correctly resulted in a score of 1 (i.e., "mastery") and performing either one or none of the tasks correctly resulted in a score of 0 (i.e., "nonmastery"). Thus, there were eight possible response vectors, {000}, {100}, {010}, etc. Assuming that normal development is consistent with the three levels of spatial tasks, the permissible response vectors comprise those of a linear scale: {000}, {100}, {110}, and {111}. Frequencies for the 573 children are shown in Table 4.1, where it is apparent that most of the observed response vectors are consistent with a linear scale (actually, 566 out of 573, or 98.8%). Preliminary to further analysis, an independence model (i.e., an unrestricted, one-class model) was fitted to these data and resulted in a very poor fit as evidenced by a G^2 value of 427.20 with $8 - 3 - 1 = 4$ degrees of freedom. Next, parameters for the four models presented in the previous section (i.e., Proctor, intrusion–omission error, variable-specific error, and latent-class-specific error) were estimated using the MLLSA program. Expected frequencies for the four models are presented in Table 4.1 and parameter estimates along with fit statistics are summarized in Table 4.2. It is notable that all four models fit the data satisfactorily in the sense that the chi-square tests of fit do not indicate significant departures. Both the variable-specific and the latent-class-specific error models illustrate a problem that arises when exceptionally good model fit occurs. That is, in both cases the estimates for some of the error rates are at the lower boundary value of 0.0. In such cases, the rank of the asymptotic variance–covariance matrix for the estimates

TABLE 4.1
Left-Right Clinical Scale

Level {ABC}	Frequency	Expected Frequencies				Modal Post. Prob.	Modal Class
		Proctor	Intr/Omis	Var-Spec	LC-Spec		
{000}*	170	169.42	169.99	170.92	170.00	0.994	1
{100}*	73	72.89	73.00	73.01	73.02	0.922	2
{010}	6	3.72	4.52	5.02	5.71	0.998	3
{110}*	254	255.40	255.45	254.05	254.13	0.997	3
{001}	0	1.49	0.08	0.00	0.01	0.784	2
{101}	1	1.22	1.20	1.90	0.97	0.867	2
{011}	0	0.61	1.18	0.00	0.13	0.999	3
{111}*	69	68.27	67.58	68.03	69.01	0.917	4
Total	573	572.99	573.00	572.93	572.98		

NOTE: The asterisks (*) denotes permissible response pattern.

TABLE 4.2
Models Fitted to Clinical Scale Data

Model	Permissible Response Pattern				Error Prob	G^2	DF	Prob.	AIC^*	BIC^*	π^*	I_D
	{000}	{100}	{110}	{111}								
I	0.302	0.124	0.455	0.118	.009	5.441	3	0.142	−0.559	−13.612	0.012	0.007
II	0.295	0.122	0.460	0.124	.0003, .017	3.028	2	0.220	−0.972	−9.674	0.012	0.005
III	0.307	0.118	0.453	0.122	.001, .027, .000*	0.854	2	0.652	−3.146	−11.848	0.012	0.003
IV	0.295	0.122	0.473	0.111	.000*, .012, .022, .000*	0.330	2	0.848	−3.670	−12.372	0.012	0.001

NOTES: Models: I. Proctor model; II. Intrusion–omission error model; III. Variable-specific error model; IV. Latent-class-specific error model. The asterisks (*) denote boundary values; degrees of freedom are adjusted accordingly.

51

will be less than expected and degrees of freedom must be adjusted accordingly (in effect, additional restrictions are being imposed on the estimates). For the variable-specific error model, one would expect the degrees of freedom to be $8 - 6 - 1 = 1$, but because the error term, $\hat{\pi}_{v3}$, is estimated as 0 and, in effect, is restricted to 0, the correct degrees of freedom are 2. Similarly, for the latent-class-specific error model, two of the estimated error rates, $\hat{\pi}_{c1}$ and $\hat{\pi}_{c4}$, are at the boundary value, 0, and the degrees of freedom are adjusted from $8 - 7 - 1 = 0$ to 2.

The error estimates for all four models are quite small. For example, the constant error rate for the Proctor model is estimated to be only .009 and, for the intrusion–omission error model, the rates are estimated to be, respectively, .0003 and .017. From examining the expected frequencies for the models (Table 4.1), it is apparent that the error rates are determined, essentially, by the six children who showed the response vector {010}. These children correctly performed the cross-midline identification, but did not correctly perform the (presumed) prerequisite task, body identification. Thus, for the variable-specific error model, the relatively large (but still small) error rate of .027 is found for the second level task. It is interesting to note that the estimated sizes of the latent classes are very similar for the four models. That is, about 30% of the children are estimated as displaying mastery of none of the three levels of tasks, about 12% are estimated to be at the first level, about 45 to 47% are estimated to be at the second level, and about 12% are estimated to have mastered all three tasks.

Whereas all four models fit the data satisfactorily, there is some arbitrariness in choosing among them. As shown in Table 4.2, the indices of dissimilarity for the four models are all extremely small (i.e., less than .01) and do not provide much guidance. Also, the index of fit, π^*, converges to the same value, $7/573 = .012$, for all models, and this simply means that a perfectly fitting latent class model can be attained in various ways by deleting the seven children who show the nonpermissible response vectors {010} and {101}. Finally, a min(AIC*) strategy favors the latent-class-specific model, whereas a min(BIC*) strategy favors the simpler Proctor model. Note that the appropriateness of these computations is somewhat questionable, because the degrees of freedom for models III and IV were adjusted upward to allow for estimated conditional probabilities that took on boundary values of 0. It is not obvious that the theory underlying the

Akaike and Bayesian indices is appropriate in this situation. However, if we adopt the latent-class-specific error model, which also had the smallest I_D value, both the first latent class, {000}, and the fourth latent class, {111}, are estimated without error, because both $\hat{\pi}_{c1}$ and $\hat{\pi}_{c4}$ are set at the boundary value 0.

Using Bayes' theorem (Equation 2.3) for the latent-class-specific error model, the children can be classified into the development types as shown in the columns labeled "Modal Post. Prob." and "Modal Class" in Table 4.1 (these columns correspond to the columns labeled "Modal P=" and "Class=" in the MLLSA output). Note that the six children who show the nonpermissible response vector {010} are classified as arising from the permissible response vector {110} and, thus, are best represented as committing an omission error for the first task. Furthermore, the one child who shows the nonpermissible response vector {101} is also classified as arising from the permissible response vector {110} and, thus, is best represented as committing an omission error for the second task as well as an intrusion error for the third task. The proportion correctly classified is equal to $(170 \times .994 + 73 \times .922 + 6 \times .998 + 254 \times .997 + 0 \times .784 + 1 \times .867 + \cdots + 0 \times .999 + 69 \times .917)/573 = .977$ and, corrected for chance based on the estimated largest latent class proportion, $\hat{\pi}_3^X = .473$, the lambda statistic is only slightly smaller, $\lambda = (.977 - .473)/(1 - .473) = .956$.

D. Extended Models

Goodman Intrinsically Unscalable Model

Goodman (1975) introduced the notion of defining scaling models in which some respondents are scalable while other respondents are "intrinsically" unscalable. The basic concept is that scalable respondents produce responses that are completely consistent (i.e., error-free) with one of the types represented by permissible response vectors, whereas intrinsically unscalable respondents produce responses in accordance with an independence model. Thus, the respondents who display one of the nonpermissible response vectors could have arisen only from the single class of intrinsically unscalable respondents. However, respondents who display one of the permissible response vectors could have arisen from the associated latent class or, by chance, from the intrinsically unscalable class. Formally, the Goodman intrinsically unscalable model can be incorporated into any given scaling model by including one additional latent class for unscalable respondents and restricting the error rates to 0 for the latent classes corresponding to the hierarchic structure. For three variables and a linear scale of the form {000}, {100}, {110}, and {111}, there would be a total of five latent classes, where the latent class proportions, π_t^X, for $t = 1, 2, 3, 4$, correspond to the permissible response vectors, and the latent class proportion, π_5^X, corresponds to the intrinsically unscalable class. In addition, the following restrictions apply to the conditional probabilities for the first four latent classes:

$$
\begin{aligned}
\pi_{11}^{\bar{A}X} = \pi_{11}^{\bar{B}X} &= \pi_{11}^{\bar{C}X} = \pi_{12}^{\bar{B}X} = \pi_{12}^{\bar{C}X} = \pi_{13}^{\bar{C}X} = \cdots \\
&= \pi_{02}^{\bar{A}X} = \pi_{03}^{\bar{A}X} = \pi_{03}^{\bar{B}X} = \pi_{04}^{\bar{A}X} = \pi_{04}^{\bar{B}X} = \pi_{04}^{\bar{C}X} = 0
\end{aligned}
\tag{4.5a}
$$

or, equivalently,

$$
\begin{aligned}
\pi_{01}^{\bar{A}X} = \pi_{01}^{\bar{B}X} &= \pi_{01}^{\bar{C}X} = \pi_{02}^{\bar{B}X} = \pi_{02}^{\bar{C}X} = \pi_{03}^{\bar{C}X} = \cdots \\
&= \pi_{12}^{\bar{A}X} = \pi_{13}^{\bar{A}X} = \pi_{13}^{\bar{B}X} = \pi_{14}^{\bar{A}X} = \pi_{14}^{\bar{B}X} = \pi_{14}^{\bar{C}X} = 1
\end{aligned}
\tag{4.5b}
$$

Finally, the conditional probabilities, $\pi_{15}^{\bar{A}X}$, $\pi_{15}^{\bar{B}X}$, and $\pi_{15}^{\bar{C}X}$, for the fifth, or intrinsically unscalable, class are unrestricted.

It is possible that there are two or more unscalable types, and it is a simple extension of the Goodman model to include additional unscalable latent classes if the number of variables is large enough to allow identifiability of the model. With four variables, for example, a linear scale with two additional unscalable classes would require estimation of seven latent class proportions (of which six are independent) and eight conditional probabilities. Thus, based on frequencies for sixteen response vectors, there would still be 1 degree of freedom remaining to assess model fit.

Dayton and Macready Extended Model

An unrealistic assumption of the Goodman model is that respondents in the scalable classes respond without error. It is possible, for example, that respondents make errors in recording answers or, in an achievement context, that they attain correct answers by guessing. To allow for these possibilities, Dayton and Macready (1980) extended the intrinsically unscalable latent class model to include various kinds of errors for the scalable types. In fact, any of the error models presented earlier, from Proctor through latent-class-specific error, can be modified to include an intrinsically unscalable class. Thus, for example, the intrusion–omission error model, with the restrictions in Equation 4.2 for the first four latent classes, can be extended to included a fifth latent class with unrestricted conditional probabilities for the variables.

E. Lazarsfeld–Stouffer Data Example

The Lazarsfeld–Stouffer questionnaire data set (Lazarsfeld, 1950), which has appeared often in the latent class modeling literature, is used to illustrate the Goodman intrinsically unscalable class model and some extended models. This sample comprised 1,000 noncommissioned officers who responded to four dichotomous (agree/disagree) statements dealing with attitudes toward the Army. The statements, in the order A, B, C, D, were intended to represent increasingly more favorable attitudes toward the Army and it seems appropriate to assume an underlying linear scale based on the permissible response vectors, {0000}, {1000}, {1100}, {1110}, and {1111}.

As summarized in Table 4.3, model I is an intrinsically unscalable class model with one unscalable class, model II is a Goodman intrinsically unscalable class model with two unscalable classes, model III is an intrusion–omission error model, and model IV is a variable-specific error model. All models were fitted to the Lazarsfeld–Stouffer data using MLLSA. Of these models, only model II provided a marginally acceptable fit, but over 70% of respondents are estimated to be in the two unscalable classes. Given this, model II was not considered to represent a satisfactory basis for scaling the responses. Although not summarized in the table, both a Proctor model and a latent-class-specific error model also resulted in poor fits. Using the same general approach described in Section 3.C, values of the fit index, π^*, were computed for each model except model II. We note that all of these values are relatively large, ranging from .141 for the variable-specific error model to .228 for the intrusion–omission error model. In addition, the indices of dissimilarity, I_D, are included in Table 4.3. Although the values for the two error models are relatively large, .087 and .077, respectively, the index is only .041 for the one unscalable class Goodman model and, in some contexts, this

TABLE 4.3
Models Fitted to Lazarsfeld–Stouffer Attitude Data

Model	Permissible Response Pattern					Intrinsically Unscalable	Error Probabilities	Intrinsically Unscalable Prob. (Yes)	G^2	DF	Prob.	AIC°	BIC°	π°	l_0
	{0000}	{1000}	{1100}	{1110}	{1111}										
I	0.050	0.011	0.000	0.079	0.188	0.672	n/a	.696, .645, .535, .255	26.533	6	0.000	14.533	−14.914	0.143	0.041
II	0.022	0.002	0.015	0.112	0.139	0.300, 0.410	n/a	.852, .744, .651, .686, .542, .491, .437, .034	3.620	1	0.057	1.620	−3.288	n/c	0.008
III	0.193	0.080	0.127	0.338	0.261	n/a	.214, .128	n/a	71.506	9	0.000	53.506	9.336	0.228	0.087
IV	0.159	0.062	0.067	0.356	0.356	n/a	.140, .181, .234, .012	n/a	43.629	7	0.000	29.629	−4.725	0.141	0.077
V	0.020	0.022	0.067	0.284	0.127	0.481	.000, .452	.975, .922, .830, .602	5.680	4	0.227	−2.350	−21.981	0.045	0.014
VI	0.178	0.049	0.080	0.196	0.148	0.330	.240, .360, .329, .003	.997, .999, .866, .619	1.702	2	0.427	−2.298	−12.114	0.012	0.009

NOTES: Models: I, one intrinsically unscalable class model; II, two intrinsically unscalable classes model; III, intrusion–omission error model; IV, variable–specific error model; V, intrusion–omission error and one intrinsically unscalable class model; VI, variable–specific error and one intrinsically unscalable class model. n/a denotes not applicable; n/c denotes not computed.

value might be deemed acceptable. However, we proceed to fit extended models by combining the single intrinsically unscalable class with both the intrusion–omission error model (model V) and with the variable-specific error model (model VI). Both of these models fit the data well in terms of chi-square goodness-of-fit tests and in terms of the fit indices, π^* and I_D. A final choice among these models is not completely clear because a min(AIC*) or min(BIC*) strategy would select model V—intrusion–omission error with an intrinsically unscalable class—by a small margin, whereas both the π^* and I_D indices favor model VI—variable-specific error with an intrinsically unscalable class. Note that this ambiguity arises because the min(AIC*) and min(BIC*) statistics take into account the complexity of the model, and the improvement in fit afforded by model VI does not outweigh the additional complexity introduced by including two more parameters in the model. In addition, although a formal hypothesis test is not appropriate for these two nonnested models, the difference in their chi-square fit statistics is only 3.948 with 2 degrees of freedom.

Given that both extended models provide attractive choices for the Lazarsfeld–Stouffer data, we can consider what substantive difference a choice between them would make. Four features are apparent: (1) the estimated size of the intrinsically unscalable class is about 30% smaller for model VI; (2) the estimated latent class proportions are similar in magnitude except for the vector {000} that includes many more estimated respondents for model V; (3) the conditional probabilities estimated for the intrinsically unscalable class are very similar for the two models; and (4) for model VI, the intrusion error rate is estimated to be essentially 0, but the omission error rate is estimated to be the relatively large value of .452. Given these considerations, it seems reasonable, if one model must be chosen, to favor model VI—variable-specific error with an intrinsically unscalable class. Table 4.4 displays expected frequencies and classifications into the latent classes based on Bayes' theorem (the columns are defined as in Section 4.B). As reported by MLLSA, the proportion of correct classifications is .621 and the chance-corrected proportion, λ, is .426. We now note some unsettling aspects of this model choice. In particular, no respondent is classified in either of the classes associated with the permissible response vectors {1000} or {1100}, and respondents in the two large classes {1110} and {1111} are classified as arising from the unscalable class despite the fact that both of

58

TABLE 4.4
Classification Results for Model V

Item {ABCD}	Frequency	Expected Frequency	Modal Class	Modal Posterior Probability
{0000}*	75	73.65	{0000}	0.781
{1000}*	69	68.35	{0000}	0.274
{0100}	55	54.85	{0000}	0.589
{1100}*	96	96.87	{1110}	0.321
{0010}	42	44.95	{0000}	0.628
{1010}	60	60.65	{1110}	0.587
{0110}	45	42.96	{1110}	0.480
{1110}*	199	198.72	IUC	0.561
{0001}	3	4.50	{1111}	0.955
{1001}	16	13.39	{1111}	0.985
{0101}	8	7.90	{1111}	0.968
{1101}	52	51.72	IUC	0.542
{0011}	10	8.88	{1111}	0.986
{1011}	25	27.13	{1111}	0.991
{0111}	16	16.33	{1111}	0.954
{1111}*	229	229.17	IUC	0.790
Total	1000	1000.02		

NOTE: The asterisks (*) denotes permissible response pattern.

these response vectors were assumed to represent scalable types. Let it suffice to conclude that the Lazarsfeld–Stouffer data represent a challenge to model in a meaningful way and that reliance on significance tests and fit indices alone does not necessarily result in the choice of models that provide satisfying interpretations.

F. Latent Distance Models

Lazarsfeld and Henry (1968) considered the problem of ordering latent classes and proposed some principles that define so-called latent distance models. In fact, the Proctor and intrusion–omission error models are restricted versions of their model. In particular, if the notions of linear scales, as previously introduced, are assumed, then for each variable in a latent distance model there are two uniquely ordered conditional probabilities for the response of 1. Assuming three variables and the four permissible response vectors for a linear scale, {000}, {100}, {110}, and {111}, there are several equality and ordering restrictions on the conditional probabilities. These restrictions, along with a simplified notation, are

$$\pi_{11}^{\bar{A}X} \leq \left[\pi_{12}^{\bar{A}X} = \pi_{13}^{\bar{A}X} = \pi_{14}^{\bar{A}X}\right] \quad \text{or} \quad a_1 \leq b_1$$

$$\left[\pi_{11}^{\bar{B}X} = \pi_{12}^{\bar{B}X}\right] \leq \left[\pi_{13}^{\bar{B}X} = \pi_{14}^{\bar{B}X}\right] \quad \text{or} \quad a_2 \leq b_2 \qquad (4.6)$$

$$\left[\pi_{11}^{\bar{C}X} = \pi_{12}^{\bar{C}X} = \pi_{13}^{\bar{C}X}\right] \leq \pi_{14}^{\bar{C}X} \quad \text{or} \quad a_3 \leq b_3$$

(where $a_1 = \pi_{11}^{\bar{A}X}$, $b_1 = \pi_{12}^{\bar{A}X} = \pi_{13}^{\bar{A}X} = \pi_{14}^{\bar{A}X}$, etc.). Note that there are exactly two conditional probabilities associated with each manifest variable and that these conditional probabilities are ordered in a manner that corresponds with the permissible response patterns. Using the simplified notation, this ordering is displayed in Table 4.5.

For the specific case of three manifest variables, the latent distance model is not identified because the number of independent parameters is nine (i.e., three latent class proportions and six conditional probabilities), but the number of response vectors is only eight. With four variables, a latent distance model has eight distinct conditional probabilities resulting from the restrictions

$$\pi_{11}^{\bar{A}X} \leq \left[\pi_{12}^{\bar{A}X} = \pi_{13}^{\bar{A}X} = \pi_{14}^{\bar{A}X} = \pi_{15}^{\bar{A}X}\right]$$

$$\left[\pi_{11}^{\bar{B}X} = \pi_{12}^{\bar{B}X}\right] \leq \left[\pi_{13}^{\bar{B}X} = \pi_{14}^{\bar{B}X} = \pi_{15}^{\bar{B}X}\right]$$

$$\left[\pi_{11}^{\bar{C}X} = \pi_{12}^{\bar{C}X} = \pi_{13}^{\bar{C}X}\right] \leq \left[\pi_{14}^{\bar{C}X} = \pi_{15}^{\bar{C}X}\right] \qquad (4.7)$$

$$\left[\pi_{11}^{\bar{D}X} = \pi_{12}^{\bar{D}X} = \pi_{13}^{\bar{D}X} = \pi_{14}^{\bar{D}X}\right] \leq \pi_{15}^{\bar{D}X}$$

TABLE 4.5
Simplified Notation Showing Restrictions for
Latent Distance Model

Permissible Response Pattern	Conditional Probability for Item		
	A	B	C
{000}	a_1	a_2	a_3
{100}	b_1	a_2	a_3
{110}	b_1	b_2	a_3
{111}	b_1	b_2	b_3

NOTE: $a_1 \leq b_1$; $a_2 \leq b_2$; $a_3 \leq b_3$

60

In addition, because five latent class proportions must be estimated, this would seem to yield 3 degrees of freedom for assessing fit. However, the four-variable model, too, is not identified because the covariance matrix is not of full rank. In fact, only six independent conditional probabilities can be estimated and, thus, two additional restrictions must be imposed to attain identifiability. The most obvious restrictions are similar to those in the variable-specific error model (Equation 4.3), but applied only to two of the variables. For example, if these restrictions are applied to variables A and D, the corresponding rows in Equation 4.7 become

$$\pi_{11}^{\bar{A}X} = \pi_{02}^{\bar{A}X} = \pi_{03}^{\bar{A}X} = \pi_{04}^{\bar{A}X} = \pi_{05}^{\bar{A}X}$$

$$\pi_{11}^{\bar{D}X} = \pi_{12}^{\bar{D}X} = \pi_{13}^{\bar{D}X} = \pi_{14}^{\bar{D}X} = \pi_{05}^{\bar{D}X}$$

The Stouffer–Toby (1951) data were presented by Lazarsfeld and Henry (1968) to illustrate the latent distance model (see Table 4.6). This data set is based on 216 responses to four questionnaire items that were designed to measure behavior in situations involving role conflict (e.g., "Your close friend is a doctor for an insurance company. He examines you when you need insurance. He finds that you are in pretty good shape but he is doubtful on one or two minor points which are difficult to diagnose. What right do you have to expect him to shade the doubts in your favor?"). Answers suggesting a response based on friendship are described as *particularistic* (coded 0), whereas more socially desirable responses are described as *universalistic* (coded 1). The items have been ordered A, B, C, and D in terms of their presumed display of the universalistic trait. The estimated conditional probabilities for the preceding constrained model, shown conforming to the pattern of Table 4.5, are

Latent Class Proportion	Permissible Vector
1: .201	{.012, .252, .363, .136}
2: .438	{.988, .252, .363, .136}
3: .102	{.988, .939, .363, .136}
4: .019	{.988, .939, .947, .136}
5: .239	{.988, .939, .947, .864}

This model fits the data exceptionally well as evidenced by a G^2 value of .922 with 5 degrees of freedom ($p = .969$). Despite this excellent fit, the frequency data in Table 4.6 show that the response vector

TABLE 4.6
Stouffer–Toby Role
Conflict Data

Item {ABCD}	Frequency
{0000}	42
{1000}	23
{0100}	6
{1100}	25
{0010}	6
{1010}	24
{0110}	7
{1110}	38
{0001}	1
{1001}	4
{0101}	1
{1101}	6
{0011}	2
{1011}	9
{0111}	2
{1111}	20
Total	216

{1010} has a large number of respondents, although it is not a permissible response vector. Good fit for this response vector was attained because the smaller of the two conditional probabilities for items B and C are relatively large (i.e., .252 and .363 versus .012 and .136 for items A and D). For purposes of comparison, a model that includes an intrinsically unscalable class was fitted to the Stouffer–Toby data. The fit of this model, with 6 degrees of freedom, is also excellent ($G^2 = .989$, $p = .986$) and shows a somewhat smaller AIC* value ($.989 - 2(6) = -11.011$) than does the latent distance model, AIC* $= .922 - 2(5) = -9.078$ (note that BIC* also favors the intrinsically unscalable model). However, the estimated latent class proportion for the intrinsically unscalable class is .681, which suggests that over two-thirds of the respondents do not belong to any of the scalable types.

A variety of other models that included the pattern {1010} among the permissible response vectors was reported for the Stouffer–Toby data by Dayton and Macready (1980). Among these models were various of the error models summarized earlier in this chapter, although none of the models provided definitively better representation of the data than those presented in this section.

G. Latent Markov Models

Panel data arise from repeatedly measuring the same observed variables over time. In the simplest case, a single dichotomous outcome is assessed at, say, V time periods and the data can be summarized in observed response vectors of the type considered throughout this chapter. For example, consider a panel of potential voters in a state election between two candidates, coded 0 and 1, that is surveyed on three occasions concerning their candidate preference. The response vector $\{0, 0, 1\}$ represents individuals who changed their choice of candidate between the second and third occasions of measurement, whereas the response vectors $\{0, 0, 0\}$ and $\{1, 1, 1\}$ represent individuals who were consistent in their choices on all three occasions. In practice, all eight different response vectors could be observed and represent varying degrees of consistency/inconsistency.

Certain hypothetical models of how change occurs in panel data can be represented in the same manner as linear scales. For example, if there is change over time in favor of candidate 1, then the permissible response vectors would be those of a "reverse" linear scale: $\{0, 0, 0\}$, $\{0, 0, 1\}$, $\{0, 1, 1\}$, and $\{1, 1, 1\}$. There is a long tradition of applying Markov chain models to the analysis of panel data of this type. Lazarsfeld and Henry (1968, Chap. 9) summarized so-called latent Markov models in which two or more Markov processes are postulated within latent subgroups of the panel. In this section, we present some elementary notions. For an introduction to the more recent literature on the topic of latent Markov models, the chapter by Langeheine (1994) is recommended. Also, Van de Pol, Langeheine, and de Jong (1991) have developed a computer program PANMARK. Another program for these models, WinLAT, developed by Collins (1997), has a modern computer interface (it is available on the Web site //methcenter.psu.edu/winLTA.html).

There are three fundamental components to a latent Markov model. Using the notation of Langeheine (1994), the first component, δ, is a vector of latent class proportions at the time of the initial measurement (e.g., $\delta' = \{0.79, 0.21\}$ indicates that 79% of the cases are members of the first latent class at time, $t = 1$). The second component is a matrix, R, of conditional probabilities for the responses to the variable being measured. For stationary Markov models, R is assumed to be constant with respect to time [e.g., assuming $R = \begin{pmatrix} .87 & .13 \\ .17 & .83 \end{pmatrix}$, the first row shows conditional probabilities for the 0 and 1 responses for the first latent class and the second row for the second latent class]. Finally, the third component is a latent transition matrix, T, containing probabilities for movement from one latent class to the other. Like R, this matrix is often assumed to be homogeneous with respect to time [e.g., assuming $T = \begin{pmatrix} .70 & .30 \\ .02 & .98 \end{pmatrix}$, the first row represents the first latent class and shows, from time t to time $t + 1$, the probability of staying in latent class 1 (i.e., .70) or moving to latent class 2 (i.e., .30), whereas the second row represents the second latent class and shows the probability of moving to latent class 1 (i.e., .02) or staying in latent class 2 (i.e., .98)].

It should be noted that the stationary latent Markov model with homogeneous transition matrices is essentially equivalent to an intrusion–omission error model for the "reverse" linear scale, $\{0, 0, 0\}$, $\{0, 0, 1\}$, $\{0, 1, 1\}$, and $\{1, 1, 1\}$. In particular, the elements in δ are equivalent to $\{\pi_1^X, 1 - \pi_1^X\}$ and the intrusion–omission errors are equivalent to the elements in R (i.e., .13 and .17 in the foregoing example). The illustrative values for δ, R, and T in the previous paragraph are from Example 1 in Langeheine (1994).

H. Other Scaling Models

Located Latent Class Model

Uebersax (1993) described a so-called located latent class model in which it is assumed that both the manifest variables and the latent classes are ordered along a single latent dimension. Assume that the manifest variables, A, B, etc. are located along a latent dimension at positions τ_a, τ_b, etc. Also assume that the T latent classes are located at the positions, β_t, in the sense that this point defines the median (i.e., fiftieth percentage point) of some associated cumulative function. A typical function for the vth variable is a logistic of the form

$$\Psi_{tv} = \frac{\exp(\alpha(\tau_v - \beta_t))}{1 + \exp(\alpha(\tau_v - \beta_t))} \tag{4.8}$$

where α is a constant discrimination, or scale, parameter (i.e., a parameter that determines the steepness of the slope of the logistic function in the vicinity of the median).[5] Note that given α and β_t for some particular latent class, the conditional probabilities for the variables, A, B, etc. are determined by substituting the values, τ_a, τ_b, etc., in Equation 4.8. Because the scale parameter, α, for the logistic functions defining the latent classes is constant, this process ensures a complete ordering of conditional probabilities. However, the nature of the underlying continuous latent variable along which both variables and classes are located is arbitrary with respect to location as well as scale, so certain restrictions must be imposed on the solution for the β_ts.

The located latent class model is somewhat beyond the scope of this volume, because the focus here is on models involving discrete classes. However, for purposes of comparison, the program LLCA (Uebersax, 1993) was used to fit a five-class model to the Stouffer–Toby data. Using the same format as for the latent distance model, the latent class proportions and conditional probabilities were estimated to be

Latent Class Proportion	Permissible Vector
1: .078	{.044, .139, .148, .469}
2: .163	{.062, .187, .199, .558}
3: .160	{.166, .409, .427, .791}
4: .433	{.242, .526, .545, .859}
5: .166	{.997, .999, .999, .999}

Note that the conditional probabilities are completely ordered with respect to both the items and the latent classes. The reader should consult Uebersax (1993) for more information concerning this model.

T-Class Mixture Model

Lindsay, Clogg, and Grego (1991) also presented a latent class model that is related to the notion of an underlying continuum. They define a T-class mixture model in which each manifest variable, A, B, etc., is located along a continuous latent variable and has an associated item response function. Like many models in item response theory, including the Rasch model (Andrich, 1988), this function may be logistic in form. In the case of the Rasch model, the probability of a positive response to the vth dichotomous, manifest variable is assumed to follow a logistic function of the form

$$\Psi_v = \frac{\exp(\phi_s - \theta_v)}{1 + \exp(\phi_s - \theta_v)} \tag{4.9}$$

where ϕ_s is the "ability" of the sth respondent and θ_v is the "difficulty" of the vth variable. Although, in theory, the Rasch model seems to incorporate a separate ability for each respondent, in fact, the sufficient statistics for purposes of estimation are the scores (i.e., counts $0, 1, 2, \ldots, V$ of positive responses) associated with response vectors, and there are only $V + 1$ distinct scores. An interesting result presented by Lindsay, Clogg, and Grego (1991) is that, given their conceptualization, the Rasch model may be derived as a constrained latent class model. In particular, for V variables, a T-class mixture requires, at most, $(V+1)/2$ latent classes to reproduce the fit to the data that is provided by a Rasch model. For example, with four dichotomous test items as for the Stouffer–Toby data, a suitably constrained three-class latent class model is exactly equivalent to a Rasch model. This three-class model is summarized as

Latent Class Proportion	Permissible Vector
1: .175	{.001, .007, .007, .025}
2: .587	{.160, .511, .492, .784}
3: .237	{.479, .834, .824, .946}

It is interesting to note that this (Rasch) model yields a G^2 goodness-of-fit statistic equal to 1.09 with 3 degrees of freedom, and these are exactly the same values resulting from fitting the located latent class model summarized previously. In addition to the Lindsay, Clogg, and Grego (1991) reference, more information concerning this interesting model can be found in Clogg (1988). Also, a computer program, PRASCH, that provides parameter estimates has been written by Grego (1993).

I. Repetitive Patterns

The principles developed in this chapter can be applied to cases in which each level of a linear scale is represented by more than a single variable. For example, there were actually six spatial tasks involved in the clinical scale example of Section 4.B. To simplify the data, we chose to score pairs of items as correct/incorrect, but the modeling could have been based on the full set of six items by defining the linear scale in terms of repetitive patterns of responses. Letting the two tasks at each level be denoted as $A1$, $A2$, $B1$, $B2$, $C1$, and $C2$, the permissible response vectors for a linear scale become {000000}, {110000}, {111100}, and {111111}. The frequency data available for analysis are for the $2^6 = 64$ response vectors, and various models, such as Proctor, intrusion–omission, etc., can be defined accordingly. In general, this technique is quite flexible because it is not necessary that each level of the linear scale be represented by the same number of variables. In practice, the main drawback to this approach is often the very large number of response vectors that result.

5. JOINT SCALES

A. Introduction

Although responses to a set of variables may display hierarchic properties, a single linear scale may not describe the complexity of the data appropriately. In this chapter, we consider joint scales that arise by combining the permissible response vectors from two or more different linear scales. There are no new principles involved in parameter estimation, significance testing, etc., so the focus is on defining and illustrating joint scales.

B. Biform and Multiform Scales

A biform scale is based on the unique set of permissible response vectors from two different linear scales. For example, with four mathematics test items, A, B, C, and D, assume that the following two linear structures are hypothesized: (1) item A is prerequisite to item B that in turn is prerequisite to item C that in turn is prerequisite to item D. (2) Item B is prerequisite to item A that in turn is prerequisite to item C that in turn is prerequisite to item D. Symbolically, the two linear scales can be represented as $A \rightarrow B \rightarrow C \rightarrow D$ and $B \rightarrow A \rightarrow C \rightarrow D$. The first linear scale is based on the permissible response vectors, {0000}, {1000}, {1100}, {1110}, and {1111}, whereas the second linear scale is based on the alternate set of permissible response vectors {0000}, {0100}, {1100}, {1110}, and {1111}.

The union of these two sets comprises the six permissible response vectors for a biform scale: {0000}, {1000}, {0100}, {1100}, {1110}, and {1111}. Note that the first four vectors contain all possible combinations of 0/1 responses to items A and B. Because these two items can be considered to be unordered, the same final set of permissible response vectors is obtained if we assert that items A and B are both prerequisite to item C that in turn is prerequisite to item D, but items A and B have no prerequisite relationship. In symbols, this structure can be represented as $A \leftrightarrow B \rightarrow C \rightarrow D$, where the double-headed arrow ($\leftrightarrow$) is read as "has no prerequisite relation with." In practice, it is often this latter conceptualization of the scales that is appropriate. Thus, if two variables can logically be ordered differently within alternate scales, then, in effect, these two variables do not exhibit prerequisite relationships.

More complex scales, referred to as multiform scales, can be formed by including additional linear scales in the structure. In addition to the linear scales $A \rightarrow B \rightarrow C \rightarrow D$ and $B \rightarrow A \rightarrow C \rightarrow D$, suppose it is assumed that a third linear scale exists: Item A is prerequisite to Item C that in turn is prerequisite to Item B that in turn is prerequisite to Item D. Symbolically, this scale is $A \rightarrow C \rightarrow B \rightarrow D$ and comprises the permissible response vectors {0000}, {1000}, {1010}, {1110}, and {1111}. The union of the three separate sets of permissible response vectors yields the seven permissible response vectors for the multiform scale: {0000}, {1000}, {0100}, {1100}, {1010}, {1110}, and {1111}.

In practice, researchers sometimes posit a linear scale for a set of variables, estimate parameters for the model, find that the linear scale does not fit the data very well, and then augment the linear scale with additional permissible response vectors to improve the fit. In this process, biform or multiform scales may arise to provide reasonable models for the data. It should be noted that in this process, the researcher has gone from a confirmatory mode of analysis, where significance tests and confidence intervals have their usual textbook interpretations, to an exploratory mode of analysis that requires greater caution. It is not clear, for example, what a .05 significance level (or a 95% confidence interval) means if the model being fitted to the data has, at least in part, been derived by noting where a linear scale provides poor fit and then augmenting the scale appropriately with additional permissible response vectors. Conventional wisdom suggests that the newly "discovered" scale should be subjected to cross-validation, although this is rarely done in practice. In this regard, we follow other conventional wisdom and urge caution when interpreting results that are exploratory in nature.

C. IEA Bus Data

The International Association for the Evaluation of Educational Achievement (IEA) has conducted worldwide achievement testing of school children for the purposes of allowing comparisons among nations (Elley, 1992). A component of the 1991 assessment of reading competency for the nine-year old cohort of school children was a series of paragraphs, along with multiple-choice test items based on the content of each paragraph. One of the paragraphs, described here as "Bus," presented children with information about a bus schedule and posed four questions based on reading and interpreting the schedule. Data for the sample of 6,359 school children that responded to the Bus items in the United States are summarized in Table 5.1. The four items, labeled A, B, C, and D, become progressively more difficult for the children; the proportion answering each item correctly is, respectively, .690, .516, .272, and .080. The average proportion of correct responses for the four items is .390, which suggests that these items were quite difficult for United States school children. Based on this observation, it seems reasonable to posit a linear scale and to fit various error models as described in Chapter 4. Although Proctor, intrusion–omission error, and latent-class-specific error models were fitted to these data, they provided relatively poor fit and are not reported here. However, greater success was achieved with the variable-specific error model (labeled "Model I" in Table 5.1).

A common problem in fitting models to very large data sets, such as the Bus data, is that significance tests tend to suggest lack of fit for models that, in purely descriptive terms, seem to provide reasonable representations for the data. In the present case, the variable-specific error model has a relatively small index of dissimilarity, 1.6%, and a π^* value of only .067, yet the G^2 chi-square value is 46.86 with 7 degrees of freedom ($p < .001$). In an attempt to develop a better-fitting model for these data, the possibility of a biform scale was investigated. The components of the Pearson chi-square goodness-of-fit statistic, labeled "Discr I" in Table 5.1, show a very large discrepancy for the response vector {0111}, although the number of cases

TABLE 5.1
Models Fitted to IEA Bus Data

Item {ABCD}	Frequency	Expected Frequencies for Model:					
		Model I	Discr I	Model II	Discr II	Model III	Discr III
{0000}	1138	1148.88	0.103	1129.75	0.060	1130.09	0.055
{1000}	1532	1532.84	0.000	1539.17	0.033	1539.80	0.040
{0100}	502	466.37	2.722	504.42	0.012	500.06	0.008
{1100}	1354	1377.27	0.393	1352.13	0.003	1354.75	0.000
{0010}	75	70.34	0.309	73.72	0.022	74.07	0.012
{1010}	200	220.67	1.936	213.28	0.827	208.71	0.363
{0110}	198	182.69	1.283	168.40	5.203	198.34	0.001
{1110}	852	851.93	0.000	870.13	0.378	845.19	0.055
{0001}	13	23.25	4.519	21.98	3.669	22.48	3.998
{1001}	43	32.21	3.615	31.49	4.207	32.34	3.514
{0101}	9	10.88	0.325	11.68	0.615	10.78	0.294
{1101}	37	35.07	0.106	37.06	0.000	36.69	0.003
{0011}	15	13.06	0.288	10.48	1.949	6.68	10.363
{1011}	59	60.38	0.032	54.13	0.438	60.92	0.061
{0111}	23	57.51	20.708	52.53	16.600	30.64	1.905
{1111}	309	275.65	4.035	288.66	1.433	307.47	0.008
Total	6359	6359.00	40.374	6359.01	35.450	6359.01	20.677
G^2		46.86		39.62		18.54	
DF		7		6		5	
p value		0.000		0.000		0.002	
I_u		0.016		0.013		0.006	
AIC*		32.858		27.620		8.541	
BIC*		−14.445		−12.926		−25.247	

NOTE: Models: I. linear scale: II. biform scale: III. augmented biform.

for this pattern is quite small (i.e., only 23 out of 6359, or less than .4%). Among response vectors with large frequencies, there are notable chi-square components for {0100} and {1111}. Because the latter of these is already a permissible response vector, a second analysis was undertaken with {0100} as an additional permissible vector. This results in fitting a biform scale based on the first two linear scales described in Section 5.B. The expected values and components of the Pearson chi-square statistic are shown in Table 5.1 as "Model II" and "Discr II," respectively. The fit of this model, in terms of the G^2 chi-square statistic and the index of dissimilarity, is only slight improved over the linear model.

The next model considered was based on the set of permissible response vectors {0000}, {1000}, {1100}, {1110}, {1111}, {0100},

{0110}, and {0111} that reflect the discrepancies noted for model I. Note that the last three vectors are for a linear scale ignoring item A. In attempting to compute parameter estimates for this model using MLLSA, the latent class proportion for the final permissible response vector, {0111} converged to 0.0. Therefore, this vector was omitted from the set of permissible response vectors and the model was estimated for seven, rather than eight, latent classes.

The fit of the seven class model (labeled "Model III" in Table 5.1) is a substantial improvement over the first two models and has an index of dissimilarity of only about .6%, although the G^2 chi-square statistic is still significant at conventional levels.

Even though further refinements of this model might be possible, we turn to interpreting model III based on the parameter estimates presented in Table 5.2 because it is the model favored, among those considered, by both the min(AIC*) and min(BIC*) strategies. It should be noted that we have fitted a fairly complex latent class model in terms of the number of latent classes (i.e., seven), but that the total number of parameters in the model has been maintained at a reasonable level because, for any number of latent classes, there are only four error rate parameters—one per test item. The latent class proportions for the two permissible response vectors {0100} and {0110} represent relatively few children (i.e., .041 and .024, respectively), but these vectors, in addition to the linear scale, were necessary to achieve reasonable fit to the data. If the sample size were smaller, these small latent classes would be less desirable, but 4.1% of the cases are actually about 261 children and 2.4% are about 153 children.

TABLE 5.2
Parameter Estimates for Model III

Permissible Pattern	Latent Class Proportions	Bus Item	Error Rate
{0000}	0.208	A	0.084
{1000}	0.268	B	0.163
{1100}	0.228	C	0.033
{1110}	0.170	D	0.019
{1111}	0.063		
{0100}	0.041		
{0110}	0.024		
Total	1.000		

TABLE 5.3
Classification Results for Model III

{ABCD}	Frequency	Expected Frequency	Modal Class	Modal Post. Prob.
{0000}*	1138	1130.09	{0000}	0.8492
{1000}*	1532	1539.80	{1000}	0.8037
{0100}*	502	500.06	{0100}	0.3752
{1100}*	1354	1354.75	{1100}	0.7759
{0010}	75	74.07	{0000}	0.4433
{1010}	200	208.71	{1110}	0.7337
{0110}*	198	198.34	{0110}	0.5489
{1110}*	852	845.19	{1110}	0.9276
{0001}	13	22.48	{0000}	0.8424
{1001}	43	32.34	{1000}	0.7554
{0101}	9	10.78	{0100}	0.3434
{1101}	37	36.69	{1100}	0.5655
{0011}	15	6.68	{1111}	0.7815
{1011}	59	60.92	{1111}	0.9327
{0111}	23	30.64	{1111}	0.8726
{1111}*	309	307.47	{1111}	0.9461
Total	6359	6359.01		

NOTE: The asterisk (°) denotes permissible response vector.

The error rates for the four Bus items are relatively small except for item *B*. Further insight into these error rates can be obtained by considering the classification results displayed in Table 5.3. As earlier, the column labeled "Modal Class" shows the permissible response vector that has the largest posterior probability based on applying Bayes' theorem. Note that the response vector {1010} is classified as arising from the permissible response vector {1110} and that both response vectors {0011} and {1011} are classified as arising from the permissible response vector {1111}. In each of these cases, an error is required on item *B* and this partially explains the relatively larger error rate for this item. The success of classification can be assessed from the proportion correctly allocated, .780, and the chance-corrected value, $\lambda = .700$.

6. MULTIPLE GROUPS ANALYSIS

Many important research questions involve comparisons among groups. For example, it could be asked whether or not the latent

structure for the cheating data presented in Chapter 3 is compa-
·rable for male and female students and/or for students in different
academic programs within the university. Similarly, the linear scale
found in Chapter 4 for left–right spatial tasks could be compared
across sex and/or ethnic groups. A naive approach to these problems
would be to fit separate latent class models to the comparison groups
and then make some judgment concerning the similarity or lack of
similarity of the resulting structures. However, in this section, a modeling approach is presented in which these comparisons are made on
a statistical basis. In brief, the preferred method involves fitting the
model of interest to the combined sample of cases and then separately to the groups being compared. The model that is fitted to the
combined sample can be considered as a constrained form of the
separate-group models. Thus, various criteria, including statistical significance tests or model comparison measures such as AIC or BIC,
can be used to decide whether or not including the grouping variable in the modeling results in a better representation of the data. A
more highly technical description of some of the topics presented in
this section, along with additional examples, can be found in Clogg
and Goodman (1984).

A. Multiple-Group Extreme-Types Model

An extreme-types model was presented in Equation 3.1 for a set
of four survey items related to academic cheating by college students. We proceed to rewrite this model to include subgroups of respondents. Let G represent a manifest grouping variable with levels
$h = 1, \ldots, H$. Then, for the sth response vector in the hth group, the
unconstrained, or heterogeneous, version of the extreme-types model
for four variables, A, B, C, and D, is written as

$$
\begin{aligned}
P(y_{sh}) = \pi_{hijkl}^{G\bar{A}\bar{B}\bar{C}\bar{D}} &= \sum_{t=1}^{T} \pi_{hijklt}^{G\bar{A}\bar{B}\bar{C}\bar{D}\bar{X}} \\
&= \pi_{h1}^{G\bar{X}} \times \pi_{hi1}^{G\bar{A}X} \times \pi_{hj1}^{G\bar{B}X} \times \pi_{hk1}^{G\bar{C}X} \times \pi_{hl1}^{G\bar{D}X} + \cdots \\
&\quad + \pi_{h2}^{G\bar{X}} \times \pi_{hi2}^{G\bar{A}X} \times \pi_{hj2}^{G\bar{B}X} \times \pi_{hk2}^{G\bar{C}X} \times \pi_{hl2}^{G\bar{D}X}
\end{aligned} \tag{6.1}
$$

where the usual restrictions apply (e.g., $\pi_{h1}^{G\bar{X}} + \pi_{h2}^{G\bar{X}} = 1$ for each
group). If the observed proportion of cases in the hth group is
P_h^G, then the unconditional probability for the sth response vector is

$P(y_s) = \pi_{ijlk}^{ABCD} = \sum_{h=1}^{H} P_h^G \times P(y_{sh})$. Note that the group proportions, P_h^G, are not written in parametric form because we consider them fixed properties of the sample that do not require estimation. In any case, the overall probability, $P(y_s)$, is not of central interest because the model is estimated and fitted to the combined set of response vectors across the various groups. That is, the observed data for, say, two groups and four dichotomous variables are the 2×2^4, or 32, response vectors, along with the associated frequencies of occurrence for these 32 response vectors, and the multiple group latent class model would be fitted to these 32 frequencies. It should be noted that the maximum likelihood estimates for the latent class proportions, $\pi_{h1}^{G\bar{X}}$, and conditional probabilities, $\pi_{hi1}^{G\bar{A}X}$, $\pi_{hj1}^{G\bar{B}X}$, etc., in the heterogeneous, multiple-group model summarized in Equation 6.1 can be found simply by computing estimates for each separate group in the manner described and illustrated in Chapter 3.

When fitting a latent class model to a single group of respondents, the expected frequencies are restricted to sum to the sample size, N. For multiple-group latent-class models, the expected frequencies have additional restrictions involving the group sample sizes. For example, with two groups of sizes N_1 and N_2, where $N = N_1 + N_2$, the expected frequencies for a model must sum to N_1 in the first group and to N_2 in the second group. These sample size restrictions must be taken into account when estimating parameters and when calculating degrees of freedom for multiple-group latent class models.

A comparison of interest in most multiple-group latent class analyses is between the heterogeneous model as represented in Equation 6.1 and a constrained model in which none of the parameters is conditioned on the grouping variable. This constrained model, known as the model of complete homogeneity, is obtained from Equation 6.1 in the case of four variables and two latent classes by imposing the restrictions

$$\pi_{11}^{G\bar{X}} = \pi_{21}^{G\bar{X}} = \cdots = \pi_{H1}^{G\bar{X}}$$

$$\pi_{1i1}^{G\bar{A}X} = \pi_{2i1}^{G\bar{A}X} = \cdots = \pi_{Hi1}^{G\bar{A}X} \qquad \pi_{1j1}^{G\bar{B}X} = \pi_{2j1}^{G\bar{B}X} = \cdots = \pi_{Hj1}^{G\bar{B}X}$$

$$\pi_{1k1}^{G\bar{C}X} = \pi_{2k1}^{G\bar{C}X} = \cdots = \pi_{Hk1}^{G\bar{C}X} \qquad \pi_{1l1}^{G\bar{D}X} = \pi_{2l1}^{G\bar{D}X} = \cdots = \pi_{Hl1}^{G\bar{D}X} \qquad (6.2)$$

$$\pi_{1i2}^{G\bar{A}X} = \pi_{2i2}^{G\bar{A}X} = \cdots = \pi_{Hi2}^{G\bar{A}X} \qquad \pi_{1j2}^{G\bar{B}X} = \pi_{2j2}^{G\bar{B}X} = \cdots = \pi_{Hj2}^{G\bar{B}X}$$

$$\pi_{1k2}^{G\bar{C}X} = \pi_{2k2}^{G\bar{C}X} = \cdots = \pi_{Hk2}^{G\bar{C}X} \qquad \pi_{1l2}^{G\bar{D}X} = \pi_{2l2}^{G\bar{D}X} = \cdots = \pi_{Hl2}^{G\bar{D}X}$$

In general, for H levels of the grouping variable and for V dichotomous variables, the number of independent parameters estimated in the heterogeneous extreme-types model is $H \times (2V + 1)$. In addition, there are H restrictions on the sums of the expected frequencies imposed by the group sample sizes. Thus, the degrees of freedom for fitting the heterogeneous model are $H \times 2^V - 2H \times V + 2H = H \times (2^V - 2V - 2)$. On the other hand, the general form of the model of complete homogeneity has a total of $2V + 1$ independent parameters, as well as the same H restrictions on the sums of the expected frequencies. Thus, the model of complete homogeneity has $H \times 2^V - (2V + H + 1)$ degrees of freedom, and the difference in degrees of freedom for the unconstrained and constrained models is equal to $2V \times (H - 1) + H - 1$. These two models are nested and the restrictions that are imposed do not involve setting parameters to boundary values of 0. Thus, the use of chi-square difference tests is legitimate in this case. Note that, in general, it is not necessary to simultaneously impose all of the restrictions in Equation 6.2. Thus, as illustrated for the subsequent example, various models that incorporate partial homogeneity can be defined and tested on the data.

The parameter estimates for a model of complete homogeneity based on a sample that has been divided into groups are exactly the same as the estimates that would be found if the comparable model were estimated from the combined group. However, it is possible for the two analyses to differ with respect to fitting the observed data. For example, with $V = 4$ variables, a model fit to the 16 cells may result in a satisfactory fit, whereas the same model (i.e., model of complete homogeneity) fitted to, say, 32 cells, where the sample is divided among male and female respondents, may not fit satisfactorily.

B. Multiple-Group Analysis of the Cheating Data

Summary data for male and female respondents for the cheating data are shown in Table 6.1. The combined sample size is 317, rather than 319, because two students failed to report sex on the cheating survey form. Estimated latent class proportions and conditional probabilities for male and female students, as well as for the combined sample, are shown in Table 6.2 (the combined group estimates differ slightly from those reported in Table 3.1 because of the slightly different samples involved). The separate-group estimates resulted from separate analyses for male and female samples using a program based

TABLE 6.1

Cheating Data for Male and Female Respondents

Item {ABCD}	Sex		Total
	Male	Female	
{0000}	99	107	206
{1000}	5	5	10
{0100}	1	11	12
{1100}	3	8	11
{0010}	1	6	7
{1010}	1	0	1
{0110}	0	1	1
{1110}	1	0	1
{0001}	18	28	46
{1001}	1	2	3
{0101}	1	3	4
{1101}	2	2	4
{0011}	2	3	5
{1011}	1	1	2
{0111}	1	1	2
{1111}	0	2	2
Total	137	180	317

on the Newton–Raphson procedure (i.e., MODEL3G, Dayton and Macready, 1977) that yields estimated standard errors (the same estimates could be obtained from LEM and, without the standard errors, from MLLSA or LCAG). The combined group analysis, based on MLLSA, treated sex as an additional manifest variable (i.e., there was a total of five variables), but fixed the conditional probabilities for sex at the values $P_1^G = 137/317 = .42318$ (i.e., the proportion of males) and $P_2^G = 180/317 = .57682$ (i.e., the proportion of females) for **both latent classes**. In addition, the set of nine restrictions implied by Equation 6.2 were imposed. Note that standard errors for the total sample were presented already in Table 3.3. Viewed purely descriptively, the estimated proportion of persistent cheaters (i.e., the first latent class) is somewhat larger for males than for females, but several of the conditional probabilities for the specific cheating behaviors are larger for females than for males.

The G^2 values reported in Table 6.2 suggest that separate-group, extreme-types models provide reasonable fits to both the male ($G^2 = 7.303$, degrees of freedom = 6, $p = .294$) and female ($G^2 = 8.660$, de-

TABLE 6.2
Two-Class Solution for Cheating Data

| | Class 1 | | | | | Class 2 | | | | |
| | Males | | Females | | | Males | | Females | | |
Item	Estimate	SE	Estimate	SE	Total	Estimate	SE	Estimate	SE	Total
A	0.475	0.202	0.644	0.321	0.571	0.015	0.040	0.020	0.045	0.017
B	0.417	0.193	0.693	0.281	0.590	0.000	0.032	0.064	0.055	0.024
C	0.270	0.146	0.188	0.104	0.215	0.000	0.020	0.059	0.021	0.037
D	0.405	0.152	0.375	0.130	0.378	0.139	0.039	0.209	0.036	0.183
LC prop	0.190	0.119	0.146	0.108	0.163	0.810		0.854		0.837
N	137		180		317					
G^2	7.303		8.660		28.992					
Degrees of freedom	6		6		21					
p-value	0.294		0.194		0.114					

grees of freedom = 6, $p = .194$) samples separately. The fit of the heterogeneous model for the combined sample of male and female students is found by summing the G^2 values and the degrees of freedom from the separate-group analyses. This yields $G^2 = 7.303 + 8.660 = 15.963$ with $6 + 6 = 12$ degrees of freedom ($p = .193$). In addition, the fit of the model of complete homogeneity to the combined sample is satisfactory ($G^2 = 28.992$, degrees of freedom = 21, $p = .114$; the data may be viewed as somewhat sparse, but the Pearson chi-square statistic, $X^2 = 25.017$ with $p = .246$, and the Read–Cressie statistic, $I^2 = 25.598$ with $p = .227$, are comparable and lead to the same conclusion). It should be noted that the chi-square value for fitting the model of complete homogeneity to the combined sample is based on expected frequencies for the 32 cells that represent the response vectors for males and females combined, and this is somewhat different than the situation reported in Chapter 3, where an extreme-types model was fitted, in effect, to the marginal distribution of the four cheating items found by summing across the male and females students. However, the analysis reported in Chapter 3 also suggested adequate fit ($G^2 = 7.764$, degrees of freedom = 6, $p = .256$).

An additional question of interest is whether or not the same latent class structure typifies male and female students. Using a chi-square difference statistic, the significance test is based on the difference in G^2 values for the unconstrained (i.e., heterogeneous) and constrained (i.e., completely homogeneous) models fitted to the combined sample. This difference is $28.992 - 15.963 = 13.029$, with degrees of freedom equal to $21 - 12 = 9$, and this value is nonsignificant ($p = .161$). This can be interpreted as meaning that the model of complete homogeneity does not fit the combined sample any worse than the heterogeneous, separate-group model and, thus, there is no interpretable difference between the latent structures for the male and female samples. An alternate method of selecting between these two models is the Akaike information criterion, AIC*. The values of AIC* for the homogeneous and heterogeneous models are $28.992 - 2(21) = -13.008$ and $15.963 - 2(12) = -8.037$, and, based on a min(AIC*) strategy, this too favors the model of complete homogeneity (comparable values for BIC* are -91.945 and -53.144).

C. Significance Tests for Individual Parameters

Some researchers would not be satisfied with the global comparison reported in the previous section, but would want to examine the cheating behaviors and estimated proportions of persistent cheaters in more detail. For two different samples, assume that estimates, $\hat{\theta}_1$ and $\hat{\theta}_2$, and associated standard errors, $S_{\hat{\theta}_1}$ and $S_{\hat{\theta}_2}$, are available for some parameter of interest, θ. A large-sample z test for equality in the corresponding populations (i.e., $H_0: \theta_1 = \theta_2$) can be set up as:

$$z = \frac{\hat{\theta}_1 - \hat{\theta}_2}{\sqrt{S_{\hat{\theta}_1}^2 + S_{\hat{\theta}_2}^2}} \qquad (6.3)$$

Whereas this is a large-sample test, critical values from the unit normal distribution are used (e.g., 1.96 for a two-tailed test at the .05 level of significance). As an example, consider the estimates, $\hat{\pi}_{111}^{G\bar{B}X} = .417$ (males, positive response to cheating behavior B, "lied to avoid turning in a term paper," in the persistent cheater latent class) and $\hat{\pi}_{211}^{G\bar{B}X} = .693$ (females, same response). To test the equality of this reported cheating behavior for male and female respondents in the persistent-cheater latent class, the appropriate test statistic is $z = (.417 - .693)/\sqrt{.193^2 + .281^2} = -.276/.341 = -.81$, a nonsignificant value. However, consider cheating behavior C, "purchased term paper/obtained copy of exam," in the noncheater class, where the conditional probabilities are estimated to be $\hat{\pi}_{1|2}^{G\bar{C}X} = .000$ and $\hat{\pi}_{2|2}^{G\bar{C}X} = .059$ for males and females, respectively. The associated z value is $z = (.000-.059)/\sqrt{.020^2 + .021^2} = -.059/.029 = -2.03$, a significant value at the .05 level. Note, however, that the conservative approach would be **not** to interpret this difference, because the overall comparison between males and females does not suggest any systematic difference between the sexes with respect to cheating behavior.

D. Partial Homogeneity

It is possible to define latent class models for two or more groups that are homogeneous in some parameters but heterogeneous in others. There are, however, some conceptual issues that should be considered when setting up such models. In particular, the conditional probabilities for the variables are the basis for interpreting the latent structure in that these probabilities characterize the various subtypes of cases. If these conditional probabilities are different for, say, male and female respondents, then it seems that the latent structure is different. However, if the conditional probabilities for the variables are equal across sexes but the proportion of cases in the latent classes are different, this can be interpreted as meaning that the latent structure, in the sense of subtypes of individuals, is the same, but that there are distributional differences among these subtypes. Thus, a model of partial homogeneity that omits the first set of restrictions from Equation 6.2 (i.e., omits equality of latent class proportions, $\pi_{11}^{GX} = \pi_{21}^{GX} = \cdots = \pi_{H1}^{GX}$) may be of interest in some applications. For the extreme-types model with V variables, this model of partial homogeneity has $H \times 2^V - 2(V + H)$ degrees of freedom and can be compared either with the model of complete homogeneous or with the heterogeneous model by means of a difference chi-square statistic (or information measures).

E. Multiple-Group Scaling Models

The scaling models presented in Chapters 4 and 5 can be explored within the context of two or more groups of respondents. In general, heterogeneous models, models of complete homogeneity, and models of partial homogeneity can be set up, estimated, and assessed for fit to data using the same principles presented previously for extreme-types models.

To demonstrate these ideas, the data for the left–right spatial tasks presented in Section 4.B were separated into male and female groups. The intrusion–omission error model for a linear scale, which was one of the models that fitted the total sample reasonably well, was assessed. The frequencies for the response vectors by sex are summarized in Table 6.3 and results are presented in Table 6.4. The columns in Table 6.4 labeled "Males" and "Females" contain estimates for latent class proportions, intrusion errors, and omission errors for an unconstrained, or heterogeneous, model whereas the column labeled "Total" is for the model of complete homogeneity in which all parameters are assumed to be equal for male and female children. Based on G^2 values, the intrusion-omission error model provides good fit for both the male and female children (i.e., values of 4.469 and 1.642 with 2 degrees of freedom, $p = 0.107$ and 0.440, respectively). The fit of a heterogeneous model is assessed by summing these G^2 values, $4.469 + 1.642 = 6.111$, and, with 4 degrees of freedom, this value indicates satisfactory fit ($p = .191$).

A model of complete homogeneity for the spatial tasks is defined by restricting both the latent class proportions and the intrusion-omission error rates to equality across the male and female children. As shown in the "Total" column of Table 6.4, the model of complete

TABLE 6.3
Spatial Data for Male and Female Respondents

Level {ABC}	Total Frequency	Male	Female
{000}*	170	82	88
{100}*	73	44	29
{010}	6	5	1
{110}*	254	105	149
{001}	0	0	0
{101}	1	0	1
{011}	0	0	0
{111}*	69	30	39
Total	573	266	307

NOTE: The asterisks (*) denote permissible response vector.

TABLE 6.4

Intrusion–Omission Error Solution for Spatial Data

Latent Class	Males Proportion	Females Proportion	Total Proportion
{000}	0.303	0.286	0.295
{100}	0.158	0.091	0.122
{110}	0.423	0.492	0.460
{111}	0.116	0.131	0.124
Intrusion error rate	0.000	0.000	0.000
Omission error rate	0.029	0.008	0.017
N	266	307	573
G^2	4.469	1.642	16.545
Degrees of freedom	2	2	10
p-value	0.107	0.440	0.085

homogeneity provides adequate fit to the data ($G^2 = 16.545$, degrees of freedom $= 10$, $p = .085$). The model of complete homogeneity is nested within the heterogeneous model and, because no latent class proportion is restricted to a boundary value, the models can be compared using the difference chi-square statistic, $G^2 = 16.545 - 6.111 = 10.434$ with $10 - 4 = 6$ degrees of freedom. This value is nonsignificant ($p = .108$) at conventional levels and, on this basis, the model of complete homogeneity would be preferred. Also, the model of complete homogeneity is chosen by a min(AIC*) or min(BIC*) strategy (the AIC* values are, respectively, -3.455 and -1.889, whereas those for BIC* are -41.044 and -16.925).

The parameter estimates for the model of complete homogeneity (Table 6.4) suggest that intrusion errors are, for all practical purposes, nonexistent (the estimate for the intrusion error rate is actually .0003, but has been rounded to .000 in the table). However, omission errors are estimated to occur about 1.7% of the time, and these errors explain, for example, the six cases showing the response vector {010}.

F. Additional Considerations

The inclusion of a grouping variable in a latent class model can permit certain analyses that would otherwise be impossible due to lack of identification. Consider, for example, the case of $V = 2$ variables.

Because there are four response vectors, only very limited modeling is possible (in fact, only the two-class Proctor model results in positive degrees of freedom). However, if the sample can be blocked into two or more groups, then there would be adequate degrees of freedom to assess the fit of a homogeneous two-class model. In particular, for two groups, there would be a total of eight response vectors with 1 degree of freedom to assess fit because since there are five independent parameters to estimate (i.e., one latent class proportion and four conditional probabilities) as well as two restrictions imposed by the sample sizes for the two groups. These notions apply to scaling models as well. For example, a variable-specific error model is saturated when fitted to three variables although there would be positive degrees of freedom for a homogeneous version of this model if the data were split between two groups. Despite this advantage of blocking, it should be noted that the resulting modeling has some very real restriction, because only homogeneous models can be assessed.

7. CONCOMITANT-VARIABLE MODELS

A. The Concomitant-Variable Latent Class Model

The latent structure for a set of variables may vary with respect to characteristics of the respondents. In Chapter 6, the latent structure was conditioned on manifest categorical groups such as males and females. In the present chapter, these notions are extended to cases in which the conditioning variable may be continuous or in which there may be some combination of categorical and continuous variables. For example, responses to the cheating survey discussed in Chapter 3 may vary according to the academic performance of students as summarized by grade point average (GPA). In this context, it seems reasonable to model the latent structure such that the probability of membership in the latent class representing persistent cheaters, for example, is a monotone decreasing function of GPA. At the same time, it may be desirable to include the sex of the student as an additional explanatory variable. In general, such models are referred to as concomitant-variable (or covariate) latent class models, and both GPA and sex, for example, could play the role of concomitant variables in a single latent class model. Because of computational limitations, only the concomitant variable version of the extreme-types model is considered here and, even for this case, the focus is on the problem of creating a functional relationship between one or more concomitant variables and the latent class proportion. In theory, the model can be extended to include other functional relationships that involve conditional probabilities for the variables (see Dayton and Macready, 1988a, b).

An extreme-types model, of the type described in Chapter 3, is written for the case of four manifest variables, A, B, C, and D, as

$$
\begin{aligned}
P(y_s) = & \ \pi_1^X \times \pi_{i1}^{\bar{A}X} \times \pi_{j1}^{\bar{B}X} \times \pi_{k1}^{\bar{C}X} \times \pi_{l1}^{\bar{D}X} \\
& + \cdots + \pi_2^X \times \pi_{i2}^{\bar{A}X} \times \pi_{j2}^{\bar{B}X} \times \pi_{k2}^{\bar{C}X} \times \pi_{l2}^{\bar{D}X}
\end{aligned}
\tag{7.1}
$$

Now, consider a generalization of this model in which the latent class proportion, π_1^X, is functionally dependent on a vector of q manifest concomitant variables, $Z = \{Z_1, Z_2, \ldots, Z_q\}$. As in multiple regression analysis, no restrictive assumption is made concerning these variables. For example, Z_1 could be a continuous variable, Z_2 could be a dummy-coded variable representing group membership, and Z_3 could be the product, $Z_1 \times Z_2$, used to represent the interaction of Z_1 and Z_2. Further, assume that the functional relationship between π_1^X and the Zs is of the form

$$
\pi_{1|Z}^X \equiv g(Z, \beta)
\tag{7.2}
$$

where $g(\cdot)$ is some specified monotonic increasing (or, decreasing) function based on the parameter vector, β. A widely used functional form has been the logistic,[6] and for the case of, say, two concomitant variables, Z_1 and Z_2, Equation 7.2 can be rewritten as

$$
\begin{aligned}
\pi_{1|Z}^X &= \frac{\exp(\beta_0 + \beta_1 Z_1 + \beta_2 Z_2)}{1 + \exp(\beta_0 + \beta_1 Z_1 + \beta_2 Z_2)} \\
&= \frac{1}{1 + \exp(-(\beta_0 + \beta_1 Z_1 + \beta_2 Z_2))}
\end{aligned}
\tag{7.3}
$$

where $\beta' = \{\beta_0, \beta_1, \beta_2\}$ are logistic regression coefficients. Combining this function with the extreme-types model in Equation 7.1 results in the concomitant-variable latent class model

$$P(\mathbf{y}_s \mid \mathbf{Z}_s) = \pi_{1|\mathbf{Z}_s}^X \times \pi_{i1}^{\bar{A}X} \times \pi_{j1}^{\bar{B}X} \times \pi_{k1}^{\bar{C}X} \times \pi_{l1}^{\bar{D}X}$$
$$+ \cdots + (1 - \pi_{1|\mathbf{Z}_s}^X) \times \pi_{i2}^{\bar{A}X} \times \pi_{j2}^{\bar{B}X} \times \pi_{k2}^{\bar{C}X} \times \pi_{l2}^{\bar{D}X} \tag{7.4}$$

where $\mathbf{y}_s$ is the response vector and $\mathbf{Z}_s$ is the vector of concomitant variables for the sth respondent. Note that the conditional probabilities in the first latent class (i.e., $\pi_{i1}^{\bar{A}X}$, $\pi_{j1}^{\bar{B}X}$, etc.) and in the second latent class (i.e., $\pi_{i2}^{\bar{A}X}$, $\pi_{j2}^{\bar{B}X}$, etc.) are independent of $\mathbf{Z}_s$. That is, the values of the conditional probabilities for the two latent classes do *not* depend upon the values of the concomitant variables.

Unfortunately, in other than the simplest cases, the data necessary to estimate parameters in concomitant variable latent class models, such as the foregoing, are rather unwieldy. If a concomitant variable takes on values over some continuum, then it is possible that the data for individual respondents are essentially unique from case to case, and computational approaches based on frequency tables are no longer feasible. In practice, it is always possible to reduce a continuous concomitant variable to a set of ordered categories and, thereby, greatly simplify the data. For example, for the cheating data analyzed in Chapter 3, each student reported their undergraduate GPA. In ungrouped form, there could be many more levels of GPA than realistically could be analyzed. However, the GPA data were collected from respondents in a set of five prescribed categories (e.g., 2.99 and less, 3.00–3.25, 3.26–3.50, etc.) and, given four items, the overall frequency table contains a manageable number of cells. That is, with $V = 4$ variables and five categories for the concomitant variable, GPA, there are $16 \times 5 = 80$ cells with a sample size of 319 (although the frequency table will be sparse).

B. Parameter Estimation

The extreme-types concomitant-variable model in Equation 7.4 contains, in general, $2 \times V$ conditional probabilities and $q + 1$ logistic function parameters for a linear model based on q concomitant variables as in the exponent of Equation 7.3. Thus, for this model, a total of $2 \times V + q + 1$ parameters must be simultaneously estimated. In theory, maximum-likelihood estimates can be found by simple extensions of the procedures described in Chapter 2 (see Dayton and Macready, 1988a). Unfortunately, none of the popular latent class programs such as MLLSA or LCAG can be used for concomitant variable latent class models. However, by means of some diligent programming within a spreadsheet such as Microsoft Excel, the analyst can derive estimates for models of reasonable size. In particular, the same nonlinear programming techniques mentioned in Section 2.C for the Rudas, Clogg, and Lindsay measure, π^*, can be used and are presented in the Web site, //www.inform.umd.edu/EDUC/Depts/EDMS. Using nonlinear programming to solve for the parameter estimates that minimize the G^2 statistic is equivalent to finding maximum-likelihood estimates. However, the distribution of G^2 may depart markedly from the corresponding, theoretical chi-square distribution because the frequency table for concomitant-variable latent class models tends to be sparse (i.e., to contain many 0 and/or small frequencies). For the purposes of assessing fit, it may be more reasonable to use the Read and Cressie (1988) I^2 statistic presented in formula 2.9 in Chapter 2.

C. Cheating Data Example

As previously noted, students reported their undergraduate GPA in five ordered categories that were used as levels of a concomitant variable to fit an extreme-types latent class model to responses to the four cheating items. The sample size for this analysis was 315, because four students failed to report information for the concomitant variable. The $16 \times 5 = 80$ cell frequency table is shown in Table 7.1. For computing purposes, each GPA category was represented by its midpoint (e.g., 3.875 for the last interval) or, in the case of the open-ended first interval, by the value 2.875. The Excel programming and detailed output that resulted from applying the concomitant-variable model is included in the Web site, //www.inform.umd.edu/EDUC/Depts/EDMS. The final estimates for the conditional probabilities for a "yes" response for the concomitant variable model are very similar to those reported in Chapter 3 for the extreme-types model. As in the earlier analysis, the first latent class can be interpreted as representing persistent cheaters. For comparison, these two sets of values are

TABLE 7.1

Cheating Data Summarized by GPA Levels

Item {ABCD}	Frequency for GPA				
	2.99 & <	3.00–3.25	3.26–3.50	3.51–3.75	3.76–4.00
{0000}	51	63	35	30	24
{1000}	5	4	0	0	1
{0100}	4	4	4	0	1
{1100}	4	6	1	0	0
{0010}	0	5	1	0	1
{1010}	1	0	0	0	0
{0110}	1	0	0	0	0
{1110}	1	0	0	0	0
{0001}	19	18	6	2	1
{1001}	2	1	0	0	0
{0101}	3	0	0	1	0
{1101}	4	0	0	0	0
{0011}	4	0	0	0	1
{1011}	0	1	1	0	0
{0111}	0	1	0	1	0
{1111}	1	1	0	0	0
Total	100	104	48	34	29

summarized:

	Extreme Types		Concomitant Variable	
Item	LC1	LC2	LC1	LC2
A	.579	.017	.561	.010
B	.591	.030	.514	.035
C	.217	.037	.215	.035
D	.377	.182	.408	.174

The equation for the logistic regression of the proportion in the first latent class on GPA is estimated to be

$$\hat{\pi}_{1|z}^{x} = \frac{\exp(8.960 - 3.370 \times GPA)}{1 + \exp(8.960 - 3.370 \times GPA)}.$$

Because the slope coefficient, β_1, is negative in sign, the function is decreasing as GPA increases (Figure 7.1). If the logistic function were plotted over the whole real line, it would have a characteristic ogival,

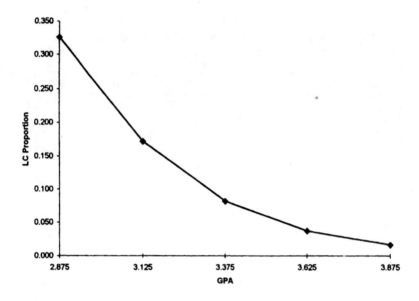

Figure 7.1. Concomitant-Variable Model for Cheating Data

or S, shape. However, for the range of GPA actually reported, only the right arm of the ogive is observed. The plot is consistent with our expectations concerning cheating among college students in that students reporting relatively high GPAs are very unlikely to be characterized as persistent cheaters, whereas it is estimated that about one-third of students in the lowest reported GPA category are persistent cheaters. If these estimated proportions are weighted by the number of students at each GPA level, the overall estimate is about 17%, which is consistent with that from the extreme-types model without a concomitant variable (i.e., 16%).

The G^2 value for the cheating data is 74.955 and the Read–Cressie I^2 value is 79.23. With 65 degrees of freedom, both values suggest reasonable fit to the data (i.e., p values of .187 and .110, respectively). Note that the degrees of freedom are computed based on the 80 cells, considering that 8 conditional probabilities and two logistic regression parameters were estimated and that the samples sizes for the five levels of GPA were fixed at 100, 104, 48, 34, and 29, respectively (i.e., 15 restrictions were imposed). Although the jackknife procedure could be used to estimate a standard error for the logistic slope coefficient, $\hat{\beta}_1$, this is very tedious because there are 80 cells in the frequency table. An alternative procedure is to restrict the slope coefficient to 0 (using the constraints in EXCEL SOLVER), reestimate the remaining parameters, and compare the chi-square fit statistics. This is equivalent to fitting an extreme-types model without the GPA concomitant variable to the data grouped by levels of GPA (note that this is different than the analysis presented in Chapter 3, because, in the previous analysis, there was no grouping). For these data, the values of G^2 and I^2 for the constrained model are, respectively, 96.609 and 86.220. Compared to the unconstrained model, these values are 21.654 and 6.984 units higher. Thus, based on 1 degree of freedom chi-squares, both of these difference statistics suggest that setting the slope to zero results in a significantly worse fit to the cheating data. In addition, the AIC* values for the unconstrained and constrained models are, respectively, −55.045 and −35.391, once again favoring the model that includes the GPA concomitant variable [a min(BIC*) strategy leads to the same conclusion based on values of −298.962 and −283.061, respectively].

D. Mixture-Binomial Concomitant-Variable Latent Class Model

The model in Equation 7.4 is much simplified if conditional probabilities are constrained to be equal within latent classes as in Section 3.G. In essence, it is assumed that a single binomial process is characteristic of each latent class. For four manifest variables, these restrictions are:

$$
\begin{aligned}
\pi_{11}^{\bar{A}X} = \pi_{11}^{\bar{B}X} = \pi_{11}^{\bar{C}X} = \pi_{11}^{\bar{D}X} \equiv \pi_{11}^{X} \\
\pi_{12}^{\bar{A}X} = \pi_{12}^{\bar{B}X} = \pi_{12}^{\bar{C}X} = \pi_{12}^{\bar{D}X} \equiv \pi_{12}^{X}
\end{aligned}
\tag{7.5}
$$

Given these restrictions, the extreme-types models can be estimated using nonlinear programming on the basis of the frequencies for the scores 0, 1, 2, 3, and 4 at each level of GPA rather than corresponding frequencies for the 16 response vectors. The estimation procedure is similar to that detailed previously and is summarized for the cheating data in the Web site, //www.inform.umd.edu/EDUC/Depts/EDMS.

The fit of the concomitant-variable latent class model is satisfactory, as indicated by both the G^2 value of 12.165 ($p = .733$ based on 16 degrees of freedom) and the I^2 value of 13.722 ($p = .619$). The

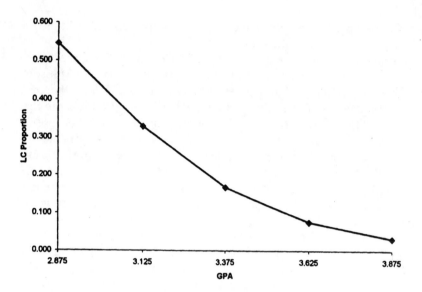

Figure 7.2. Mixture–Binomial Latent Class Model for Cheating Data

conditional probability for a "yes" response for the first latent class is
.312, which may be considered an average rate of reported cheating
behavior, although this value is considerably lower than the average,
.425, for the four items reported in Section 7.C. Similarly, the value
for the second latent class, .042, is lower than the average of those
reported in Section 7.C (i.e., .064). The plot of the logistic regression
for the binomial-mixture model (Figure 7.2) is similar in form to that
for the concomitant-variable model, although the estimated proportion of persistent cheaters is considerably higher for the lower values
of GPA. If these estimated proportions are weighted by the number of
students at each GPA level, the overall estimate is about 32%, which is
considerably higher than from either the mixture-multinomial model
(e.g., 18%) or from the extreme-types model without a concomitant
variable (i.e., 16%). In summary, although this model is somewhat
easier to fit to data, in the present case the resulting model is less
satisfactory than the more complex mixture-multinomial model.

NOTES

1. It should be noted that this simple, didactic example is based on a model that is not identified in the sense explained in Section 2.B. Therefore, if the data were actually analyzed using a latent class analysis program, the specified latent class structure could not be recovered.

2. A package of microcomputer programs, CDAS, that includes MLLSA is available from Scott R. Eliason, Department of Sociology, The University of Iowa, Iowa City, Iowa 52242; e-mail: scott-eliason@uiowa.edu. The version of MLLSA used in this book was adapted from the original mainframe computer version in 1986 by Randall Knack, a graduate student in the Department of Measurement, Statistics and Evaluation at the University of Maryland.

3. This raises a complex and controversial topic that has to do with issues related to statistical significance versus practical significance. When large samples are used in analyses, relatively small effects will, with high probability, be detected. It is, then, a practical issue as to how important these effects are. For example, very high intake of certain food products may result in measurable increases in some cancer rates, but these increases may be, relatively speaking, of minor concern given that other factors are much more important in causing the same diseases.

4. Although, in theory, there are several ways in which inequality restrictions could be imposed during maximum-likelihood estimation, available latent class analysis programs do not include this option. Nonlinear programming procedures, similar to those described in Section 3.C in the context calculating the Rudas, Clogg, and Lindsay π^* measure, could be used, because procedures such as the Excel SOLVER include provisions for imposing inequality restrictions.

5. Uebersax (1993) wrote the exponent in both the numerator and denominator of Equation 4.8 in the alternate form, $\exp(1.7\alpha(\tau_r - \beta_i))$. The constant, 1.7, changes the scaling of the parameters (e.g., α in Equation 4.8 could be reduced by a factor of 1.7 and the product, 1.7α, would remain exactly the same). This factor is included in the exponent because the resulting logistic function closely mimics the cumulative density function, or ogive, for the unit normal distribution, but does not have any substantive effect on the model or analysis. In fact, the constant α could, similarly, be absorbed into the scaling of the τ_r and β_i parameters as is done in the Rasch model.

6. The logistic function is the basis for the Rasch model in item response theory (Andrich, 1988), but other monotone functions, such as the one-parameter cumulative exponential, $\beta \times e^{-\beta Z}$, have been suggested for the concomitant-variable latent class model (Dayton and Macready, 1988b).

89

REFERENCES

Airasian, P. W. (1969) "Formative evaluation instruments: A construction and validation to evaluate learning over short time periods." Doctoral dissertation, University of Chicago.

Akaike, H. (1973). "Information theory and an extension of the maximum likelihood principle." In B. N. Petrov and F. Csáki, (eds.), 2nd International Symposium on Information Theory, Budapest: Akademiai Kiádo, pp. 267–281. Reprinted in Kotz, S., Johnson, N. L., (eds.) Breakthroughs in Statistics. Volume I: Foundations and Basic Theory. New York: Springer-Verlag.

Akaike, H. (1987) "Factor analysis and AIC." Psychometrika, 52, 317–332.

Andrich, (1988) Rasch Models for Measurement. Thousand Oaks, CA: Sage.

Bartholomew, D. J. (1987) Latent Variable Models and Factor Analysis. London: Griffin.

Bolesta, M. S. (1998) "Comparison of standard errors within a latent class framework using resampling and newton techniques." Doctoral dissertation, Department of Measurement, Statistics, and Evaluation, University of Maryland.

Clogg, C. C. (1977) Unrestricted and Restricted Maximum Likelihood Latent Structure Analysis: A Manual for Users. University Park, PA: Population Issues Research Office, Pennsylvania State University.

Clogg, C. C. (1988) "Latent class models for measuring." In R. Langeheine and J. Rost (eds.), Latent Trait and Latent Class Models. New York: Plenum Press.

Clogg, C. C. (1995) "Latent class models." In G. Arminger, C. C. Clogg, and M. E. Sobel (eds.), Handbook of Statistical Modeling for the Social and Behavioral Sciences. New York: Plenum Press.

Clogg, C. C., and Goodman, L. A. (1984) "Latent structure analysis of a set of multidimensional contingency tables." Journal of the American Statistical Association, 79, 762–771.

Collins, L. (1997) Latent Transition Analysis for Windows. University Park, PA: Pennsylvania State University.

Dayton, C. M. (1991) "Educational applications of latent class analysis." Measurement and Evaluation in Counseling and Development, 24, 131–141.

Dayton, C. M., and Macready, G. B. (1976) "A probabilistic model for the validation of behavioral hierarchies." Psychometrika, 41, 189–204.

Dayton, C. M., and Macready, G. B. (1977) "Model 3G and Model5: Programs for the analysis of dichotomous, hierarchic structures." Applied Psychological Measurement, 1, 412.

Dayton, C. M., and Macready, G. B. (1980) "A scaling model with response errors and intrinsically unscalable respondents." Psychometrika, 45, 343–356.

Dayton, C. M., and Macready, G. B. (1988a) "Concomitant-variable latent-class models." Journal of the American Statistical Association, 83, 173–178.

Dayton, C. M., and Macready, G. B. (1988b) "A latent class covariate model with applications to criterion-referenced testing." In R. Langeheine and J. Rost (eds.), Latent Trait and Latent Class Models. New York: Plenum Press.

92

Dayton, C. M., and Scheers, N. J. (1997) "Latent class analysis of survey data dealing with academic dishonesty." In J. Röst and R. Langeheine, (eds.), *Applications of Latent Trait and Latent Class Models in the Social Sciences.* Munich: Waxman Verlag.

Efron, B., and Gong, G. (1983) "A leisurely look at the bootstrap, the jackknife, and cross-validation." *The American Statistician, 37,* 36–48.

Elley, W. B. (1992) *How in the World do Students Read? IEA Study of Reading Literacy.* The Hague: IEA.

Everitt, B. S., and Hand, D. J. (1981) *Finite Mixture Models.* New York: Chapman and Hall.

Goodman, L. A. (1974) "Exploratory latent structure analysis using both identifiable and unidentifiable models." *Biometrika, 61,* 215–231.

Goodman, L. A. (1975). "A new model for scaling response patterns: An application of the quasi-independence concept." *Journal of the American Statistical Association, 70,* 755–768.

Goodman, L. A., and Kruskall, W. H. (1954) "Measures of association for cross-classifications." *Journal of the American Statistical Association, 49,* 732–764.

Grego, J. M. (1993) "PRASCH: An Fortran program for latent class polytomous response Rasch models." *Applied Psychological Measurement, 17,* 238.

Guttman, L. (1947) "On Festinger's evaluation of scale analysis." *Psychological Bulletin, 44,* 451–465.

Haberman, S. J. (1979) *Analysis of Qualitative Data, Volume 2: New Developments.* New York: Academic Press.

Haberman, S. J. (1988) "A stabilized Newton–Raphson algorithm for loglinear models for frequency tables derived by indirect observation." In C. C. Clogg (ed.), *Sociological Methodology 1988.* Washington, D. C.: American Sociological Association.

Hagenaars, J. A. (1990) *Categorical Longitudinal Analysis.* Thousand Oaks, CA: Sage.

Hagenaars, J., and Luijkx, R. (1987) "LCAG: Latent class models and other loglinear models with latent variables." Department of Sociology, Tilburg University, The Netherlands.

Kass, R. E., and Raftery, A. E. (1995) "Bayes factors." *Journal of the American Statistical Association, 90,* 773–795.

Langeheine, R. (1994) "Latent variables Markov models." In A. Von Eye and C. C. Clogg (eds.). *Latent Variables Analysis.* Thousand Oaks, CA: Sage.

Lazarsfeld, P. F. (1950) "The logical and mathematical foundation of latent structure analysis." In S. A. Stouffer et al. (eds.), *Measurement and Prediction.* Princeton, NJ: Princeton University Press.

Lazarsfeld, P. F., and Henry, N. W. (1968) *Latent Structure Analysis.* Boston: Houghton Mifflin.

Lin, T. H., and Dayton, C. M. (1997) "Model-selection information criteria for non-nested latent class models. *Journal of Educational and Behavioral Statistics, 22,* 249–264.

Lindsay, B., Clogg, C. C., and Grego, J. M. (1991) "Semi-parametric estimation in the Rasch model and related exponential response models, including a simple latent class model for item analysis," *Journal of the American Statistical Association, 86,* 96–107.

McCutcheon, A. L. (1987) *Latent Class Analysis.* Sage University Papers Series on Quantitative Applications in the Social Sciences, 07-64. Thousand Oaks, CA: Sage.

Mokken, R. J., and Lewis, C. (1982) "A nonparametric approach to the analysis of dichotomous item responses." *Applied Psychological Measurement, 6,* 417–430.